From Fear to Hate

Foundations in Language and Law

Editors
Janet Giltrow
Dieter Stein

Volume 6

From Fear to Hate

―

Legal-Linguistic Perspectives on Migration

Edited by
Victoria Guillén-Nieto, Antonio Doval Pais and Dieter Stein

ISBN 978-3-11-221418-3
e-ISBN (PDF) 978-3-11-078915-7
e-ISBN (EPUB) 978-3-11-078919-5
ISSN 2627-3950

Library of Congress Control Number: 2023935133

Bibliographic information published by the Deutsche Nationalbibliothek
The Deutsche Nationalbibliothek lists this publication in the Deutsche Nationalbibliografie;
detailed bibliographic data are available on the internet at http://dnb.dnb.de.

© 2025 Walter de Gruyter GmbH, Berlin/Boston
This volume is text- and page-identical with the hardback published in 2023.
Cover image: kokouu/E+/Getty Images
Typesetting: Integra Software Services Pvt. Ltd.
Printing and binding: CPI books GmbH, Leck

www.degruyter.com

Acknowledgements

This volume has profited from the useful insights given by colleagues at all stages of its origination, starting from comments during the sessions at the 5th ILLA General Conference in Alicante in 2021. In the final stage, it has had the congenial copyediting and comments from Donato Mancini.

Contents

Acknowledgements —— V

Victoria Guillén-Nieto, Antonio Doval Pais and Dieter Stein
Chapter 1
Emotion, language and law —— 1

Dyango Bonsignore
Chapter 2
The dangerous individual in a time of migration: *Periculum*, *dominium* and dangerousness —— 23

Monika S. Schmid
Chapter 3
Language attrition as a problem for language analysis for the determination of origin —— 49

Laura M. Hartwell
Chapter 4
"Once an alien has passed through our gates": Noncitizens in three US Supreme Court oral arguments —— 63

Mary C. Lavissière and Rachel C. Hill
Chapter 5
Improper entry by an alien on trial: the uncomfortable linguistic past of 8 United States Code § 1325 & 1326 —— 85

Jūratė Ruzaitė
Chapter 6
Impoliteness categories in hateful online comments targeting migrants in Lithuania —— 117

Fabienne Baider
Chapter 7
Covert Islamophobia and anti-Semitism via conspiracy theory —— 149

Victoria Guillén-Nieto
Chapter 8
The wording of hate speech prohibition: "You can't see the wood for the trees" —— 173

Antonio Doval Pais
Chapter 9
When the wording of the law is not enough: Hate speech crimes in Spain —— 201

Subject index —— 231

Victoria Guillén-Nieto, Antonio Doval Pais and Dieter Stein
Chapter 1
Emotion, language and law

1 Emotion and migration

Migration has become a global phenomenon, associated with social conflict and political instability. The associated problems, caused either by migration or resulting from migration, have come to dominate political discourse, giving rise to national and international laws on migration, including discriminatory provisions, pressing laws against migrants and refugees, encouraging international cooperation on migration, while fomenting antimigration rhetoric.

The present volume, *From fear to hate: Legal-linguistic perspectives on migration*, foregrounds the varying and complex emotional circumstances that migration entails. The volume title reflects the emotional process—from fear to hate—migrants and refugees undergo when they move to another country. The general aim of the volume is to analyse legal-linguistic problems at the heart of migration and to provide useful insights in pursuit of concrete solutions.

Mackenzie and Alba-Juez point to an "emotional turn" (2019: 3) in discourse analysis inspired by White's *Appraisal theory* (2002), where emotion is found in the subsystem of *affect* within the superordinate system of *attitude*. Affect is not considered a simple outward expressive behaviour, but a type of evaluation that complements rational thought. Alba-Juez and Thompson also refer to affect as a dynamic subsystem associated with the speaker's attitude or stance in discourse: "a dynamic subsystem of language permeating all linguistic levels and involving the expression of the speaker's or writer's attitude or stance towards, viewpoint on, or feelings about the entities or propositions that s/he is talking about" (Alba-Juez & Thompson 2014: 13).

When migration is embarked upon as an adventurous enterprise, in pursuit of job opportunities or better living conditions and salary, systems or bundles[1] of

---

[1] Mackenzie and Alba-Juez explain that "As a process, emotion involves different stages and facets that in discourse are better described as systems or bundles than as single, basic emotions. The systems include all the features and variables that contribute to the expression of the emotion at both the production and reception ends, including the appraisal of the situation, the expectations

Victoria Guillén-Nieto, University of Alicante
Antonio Doval Pais, University of Alicante
Dieter Stein, Heinrich-Heine-University Düsseldorf

https://doi.org/10.1515/9783110789157-001

positive emotions related to eagerness and excitement are activated. The migrant's attitude is likely to be optimistic, in anticipation of the benefits entailed by relocation to another country. In other words, the potential benefits for the migrant loom larger than the risks they must run. By contrast, when migration is a flight response to a dangerous situation associated with extreme poverty, hunger, violence, war or genocide, the emotional systems of fear and anger are activated. In these cases, the migrant's attitude is likely pessimistic, because they are forced to leave home and know that adversity and dangerous circumstances lie ahead. Here the risks the migrant must run loom larger, emotionally, than any benefits they may obtain by migration—especially since these benefits are often reduced to mere survival. Upon arrival at the borders of other countries, the odyssey for migrants and refugees often takes tragic new forms. Massive arrivals of migrants and refugees can be perceived by locals as dangerous and threatening and, therefore, can trigger negative emotions such as fear, anger or hate. These negative emotions can manifest themselves in antimigration discourse and in various forms of hate speech.

2 Emotion in antimigration discourse

Antimigration discourse is driven by negative emotions associated with fear, anger and hate. Such emotions are synthesised in Matsuda's (1989) concept of *hate speech*. Hate speech does not necessarily express hatred, as it can express negative emotions other than hatred, but it is discourse triggered by hatred. Antimigration discourse results from a negative appraisal, based on prejudice, stereotypical knowledge and intolerance, of the new and unexpected social circumstances introduced by migrants and refugees. Antimigration discourse often contains hate speech and is specifically designed to shame migrants and refugees, keep them under constant threat and, thereby, ensure that they cannot access mainstream society nor enjoy full citizenship. As Matsuda argues, "The threat of hate groups like the Ku Klux Klan and the neo-Nazi skinheads goes beyond their repeated acts of illegal violence. Their presence and active dissemination of racist propaganda mean that citizens are denied personal security and liberty as they go about their daily lives" (Matsuda 2018: 2).

The embryo of antimigration discourse lies in the representation of migration as a *crisis* (Monnier, Boursier & Seoane 2022b: 1–17) or a *threat* (Kopytowska & Chilton 2018: 133–161). Drawing on neuroscientific research into the role of language as

---

related to the emotions felt by the speaker and perceived or caused in the hearer or audience, the common-ground knowledge of the interlocutors, the gestures or emotive bodily behaviour, the polarity or valence of the emotion, etc." (2019: 16).

a fear activator, Kopytowska and Chilton analyse the interface of fear and threat in public discourse and identify the lexical, grammatical and discursive features contributing to a fear-inciting discourse whose role is to stimulate an alarm reaction in the brain (cf. Wodak [2015] 2021). Typically, migrants are portrayed as savages and invaders of other people's land. The representation of migration as a crisis or a threat in literature, news media and online (cf. Assimakopoulos, Baider & Millar 2017) has a cumulative effect. It gradually fosters and shapes negative emotions and attitudes towards migrants and refugees. Once the representation of migration as a crisis or a threat is installed in the locals' minds and integrated into their mental programming, antimigration discourse adopts a self-victimising stance, polarising society into "US vs THEM" (Schröter 2022: 83–114). The next move in antimigration discourse is to discredit migrants and refugees through fallacies, insult and disparagement, including degrading racial caricatures and dehumanising portrayals of migrants as animal-like. However simplistic hate propaganda may be, the hate-advocating speaker knows that propaganda has great power of persuasion. In other words, the hate-advocating speaker's rhetoric appeals to peoples' emotions (*pathos*) rather than to their reason (*logos*). Its emotional appeal facilitates the dissemination of hate propaganda and hate advocacy, clearing the way for incitement to discrimination, hatred, hostility, and even violence against migrants and refugees.

Hate propaganda permeates all communication genres (Guillén-Nieto 2023: 95–108), and its effects are cumulative, eventually triggering violence towards migrants and refugees. Hatred can be conveyed in a variety of ways, through covert disparate treatment and sanitised racist comments, or through overt disparate treatment, insults and outright threats of violence. According to Matsuda (2018: 23), the preference for one or another form of hate speech often depends on the speaker's socio-economic background and level of education. Speakers of lower socio-economic background and education tend to prefer more overt forms of hatred and physical violence against migrants and refugees. By contrast, speakers from higher backgrounds and with advanced educations will prefer covert, clandestine forms of hatred and symbolic forms of violence, such as righteous indignation against diversity or positive discrimination.

3 The harms of hate speech

Hate speech can hit the emotional places where migrants feel most vulnerable. Matsuda (2018: 24), Delgado and Stefancic (2004), and Delgado and Stefancic (2018) provide a thorough overview of the harms caused by hate speech. Hate speech can cause physiological effects (cf. Cannon [1915] 1925) from *emotional stress*. Such short-term harms can include but are not limited to anxiety, rapid pulse rate, hypertension and difficulty in breathing, tightening of muscles, adrenaline rushes, nightmares, and the triggering of post-traumatic stress disorder symptoms. Apart from *psychic destruction*, hate speech can devastate the migrant's *self-esteem* and sense of *personal security*. Some victims may take refuge in alcohol, drugs or other self-defeating escapes. A feeling of *shame* and *inferiority* may also be implanted in the minds of migrants and refugees after repeated exposure to hate propaganda (Matsuda 2018: 25). *Limitation of personal freedom* is another likely effect. To avoid confrontation and conflict, migrants may stay away from public places, curtail their own speech, and otherwise modify their behaviour. The response to hateful messages is internalised, their anger often repressed because direct confrontation with a hate-advocating speaker could be detrimental or even catastrophic. Lastly, hate speech can also worsen the victims' economic situation by hindering their social promotion and life chances. The victim has to bear the effects of *stereotype threat* (Delgado & Stefancic 2004: 15). Because of social prejudice and intolerance, for example, migrants and refugees may have difficulty finding jobs commensurate with their educational background and experience, or obtaining loans to buy property.

Apart from the harm done to migrants and refugees, hate speech can also have devastating effects on the hate-advocating speaker and society at large.[2] On the one hand, hate-advocating speakers can develop paranoia about the group they routinely disparage (Delgado & Stefancic 2004: 16). On the other hand, society suffers when migrants and refugees are discriminated against or otherwise abused. However, the most serious harm comes when hate speech goes unpunished, because hate speech represents a breach of constitutionally-protected democratic values and fundamental rights such as dignity, equality and respect.

4 Legal aspects

From a legal standpoint, migratory movements have always implied the need for states receiving foreign populations to regulate the migrants' conditions of stay in the country and the consequences of their non-compliance (no access to certain

---

2 Delgado and Stefancic argue that "the more diffuse kind [of hate speech]—for example, 'All niggers are inferior and should go back to Africa'—is apt to be more harmful to society in general. The more targeted variety [face-to-face hate speech]—'You goddamn nigger, go back to Africa'—harms society as well, particularly cumulatively, but its principal impact is felt by the individual victim" (2004: 12).

rights or even sanctions). This regulation can be considered "normal" in a country given "xenophobic common sense" (Ferrajoli 2019: 184), as it is aimed at basic conditions of social participation, such as access to work and public services or positions, conditions for the validity of legal acts or requirements for participation in political life.

However, migratory movements, and in particular the presence of foreign nationals in other countries, have also made it necessary for states to protect these persons against hostile reactions from certain citizens in the destination country. In this respect, the 1960s marked a turning point in the protection of minority groups. In particular, two significant international legal standards were approved and entered into force: the International Convention on the Elimination of All Forms of Racial Discrimination (1965) and the International Covenant on Civil and Political Rights (1966). These legal standards provide recommendations to states that they can adapt to their own legal systems and cultures. The legal enforcement of international standards for the prohibition of hate speech implies that, at an international level, the law recognises the social nature of emotional expression (Hess & Hareli 2019) and recognises the need to regulate emotional expression when it can cause harm, as is the case with hate speech.

So far, ordinary criminal legislation designed to protect national citizens (against, for example, insults, injuries or homicides) has not been sufficient to achieve satisfactory protection of migrants from the harmful effects of hate speech. Rather, it has been necessary to provide migrants and refugees with special protections according to their needs, which are not the same as those of national citizens (Steiker 1999: 1858). For this reason, these rules are not intended to protect all foreigners, but specifically those minority groups whose circumstances leave them vulnerable to the most serious attacks. These particular legal norms aim to protect foreign nationals who are the object of racist or xenophobic prejudice. That is, this legislation seeks compensation for their particular vulnerability. The legal field that interests us in this volume is the set of rules aimed at protecting foreign nationals from discriminatory attacks.

Essentially, two models of protection are applied by the national criminal laws, and sometimes they appear together. The first is the aggravation of liability for crimes committed on discriminatory grounds. The second is the punishment of conduct aimed at instigating, promoting or encouraging hatred, discrimination or violence. Although the expression "hate crimes" is sometimes strictly understood as only referring to crimes committed for discriminatory reasons, in a broader sense "hate crimes" could be understood to refer to both these crimes and to those that consist of instigating, promoting or encouraging hatred, discrimination or violence (hate speech crimes).

Moreover, in the specialised literature, a distinction is made between a *selective discrimination* model and an *animus* (or hostility) model, depending on what hate crime legislation requires: a discrimination effect or an act triggered by hostility or hatred, respectively. The differences are important. While in the latter it is legally irrelevant that the motivating factor does not lie with the victim (for instance, the perpetrator mistook the victim for a foreigner), this is not the case in the former model.

In either model, punishment depends on a subjective element, which, in practice, is problematic. First, in the case of aggravation of liability, it raises the question of why a crime should be punished more severely because it has been perpetrated for certain reasons. As subjective elements, the reasons for acting do not produce greater harm than the crime committed (the attack on life, physical integrity, honour, and so on, is exactly the same). Moreover, with regard to the criminal punishment of hate speech, it is difficult to explain the reasons for punishing the promotion of hatred or discrimination, while hate, in itself, or even discrimination, does not qualify as a crime. Second, proving the motives that led to the act can be challenging for the court. These difficulties have led to a discussion about whether motives are discoverable or known (as descriptivist or cognitivist theses argue) or imputable (as ascriptivist or non-cognitivist theses argue) (González Lagier 2004: 33–35).

However, the problems do not end here. In both models of criminal law intervention, numerous questions arise from various defects in the formulation of the legal norms. These defects condition the scope of criminal punishment in practice, since certain fundamental aspects—such as the type of conduct which constitutes the crime or the aggravating circumstance, the biased motives or the protected groups, or even whether these groups should only be minority groups—depend on the wording of the law. For all of the reasons above, judges' interpretation of these norms can vary significantly. Even the scope of criminal intervention often exceeds the requirement that criminal law should serve in constitutional states as the *ultima ratio legis*.

The defects affecting criminal anti-discrimination laws have grave consequences, not only for the equal treatment of citizens under criminal law but also from the point of view of respect for freedom of expression, since excessive intervention entails illegitimate curtailment. The seriousness of this can be fully appreciated if one considers the fundamental value of freedom of expression in our societies.

As the Ibero-American Court of Human Rights has emphasised, freedom of expression fulfils the dual function of protecting the right not only of a citizen to express his or her ideas or opinions, but also that of all citizens to receive information and ideas.[3] Thus, in addition to the right of each individual to freely express their

---

3 Advisory opinion ("Opinión consultiva") OC-05/85, p. 9. https://www.corteidh.or.cr (accessed 15 January 2023).

ideas or opinions, freedom of expression plays an essential role in shaping public opinion and preserving political pluralism. All international normative instruments regarding human rights and European constitutional texts recognise freedom of expression as a fundamental element. Because offences constitute criminal limits to freedom of expression, it is also essential that criminal intervention against speech should always be fully justified. However, the broad regulation of hate speech in national legislations has allowed individuals and groups that are not in need of special protection (such as politicians or members of the security forces) to criminally prosecute those who insult them on social networks.[4]

Excessive protection is absolutely counterproductive, both because it can produce a chilling effect and, consequently, inhibit certain manifestations (impoverishing participation in the public debate of ideas by silencing the voice of certain ideological positions) and because it can lead to a loss of citizens' confidence in the law and the institutions of justice. Moreover, excessive punitive intervention is unnecessary and leads to illegitimate penalties. This is intolerable in a constitutional system.

So-called "hate speech" belongs to an area outside the limits of the legitimate exercise of freedom of expression. However, while "hate speech" consists of direct provocations motivated by hatred, it lacks a precise delineation. Freedom of expression covers information and ideas that can offend, shock or disturb, as noted by the European Court of Human Rights.[5] Judges have to weigh whether a provocative utterance falls within the category of "hate speech", as its boundaries are not neatly defined by national or international legal standards.

It should be noted that the seriousness of the situations to which hate speech can lead (disrespect, dehumanisation and neglect of people, which may pave the way for acts of violence) does not justify a severe legal reaction, regardless of the seriousness of the utterance itself. The law must stand firm in the face of this temptation and must not enter into the same irrationality that lies behind those attitudes stemming from assumptions and fears that find their expression in hate speech. The law must be used with great care not to provoke disaffection among those citizens whose institutional attachment is often already weakened because they see themselves as victims rather than offenders.

At the heart of the legal problem is the balance the court must find between two opposing parties who hold competing rights. On the one hand, the hate-advocating speaker, typically affiliated with a dominant group, claims the right to freedom of expression and opinion. On the other hand, the victim, typically a

---

[4] With regard to Spain see Barata (2020: 42).
[5] Among other judgements, see *Erbakan v. Turkey*, Application No. 59405/00, 6th July 2006.

member of a minority group, claims the right to dignity. Whereas in European Law, courts must strive to find the right balance between such competing interests, under United States Constitutional Law, *inflammatory speech* cannot be prohibited unless it is "directed to inciting or producing imminent lawless action and is likely to incite or produce such action" (*Brandenburg v. Ohio* 1969). Hence, it may be reasonable to argue that the law protects society from breach of the peace in the United States but somewhat disregards the harm caused to the individual's dignity, equality and respect. By contrast, European Law prohibits incitement to discrimination, hatred, hostility or violence against legally-protected groups, and the imminence requirement is not part of the law's wording.

However, not all solutions should be legal ones. As academics insistently point out, complex problems such as this require an effective combination of social measures which favour equal basic resources for all citizens—necessary for a life with dignity with social inclusion, education and the promotion of human rights—and legal instruments of a non-punitive nature (that is, private and administrative law). Finally, criminal law should be reserved for only the most unacceptable behaviours, including certain acts that endanger life or people's physical integrity or freedom. Failure to obtain legal protection against hate speech may result in negative emotions such as impotent despair, aloneness and unfairness. In such cases, Matsuda argues, the victim becomes a "stateless person" (Matsuda 2018: 24). As mentioned, when hate speech goes unpunished, the most serious harm is caused to society itself, because the perpetrator's impunity impacts the victim's human rights and, therefore, impugns the democratic values of modern societies.[6]

The law must, therefore, show firmness when faced with the most serious acts, but it must not overstep its role. It must resort to solutions that promote, above all, social stability instead of tension. Rather than irrational authoritarianism, its solutions must arise from a nuanced understanding of the problems.

Moreover, states should not tarnish the image of migrants by considering irregular immigration illegal by itself, nor by establishing aggressive political measures that disregard the human dimension at stake in the movement of migrants and refugees. Neither should states put aside the civilising principles that all constitutional states should strive for. Such civilising principles are set aside when, for example, states establish barriers within the migrants' countries of origin in exchange for economic or other benefits, and with the implementation of pushbacks (or express deportations), unjustifiably onerous identification requirements, irregular repatriations, the ill treatment of unaccompanied minors, and so on. As Ferrajoli points out, legal inequality is a factor of miseducation in that it

---

6 For an elaborated discussion on these topics, see Chapter 8 in this volume.

generates the image of the Other as a naturally inferior being (Ferrajoli 2019: 186). We concur with Ferrajoli that it is not realistic to continue insisting on a juridical utopia, one which holds that "the pressure of the excluded on our borders can be confronted by laws and that closed borders can coexist with a future of peace" (Ferrajoli 2019: 190). The only realistic alternative is to entirely reform all laws of borders, citizenship and migration, to achieve world citizenship for all, based on the fundamental equality of human beings (Ferrajoli 2019: 184, 190–191; Balibar 2006: 13).

From the point of view of linguistics, the emotions and verbal actions triggered by and related to hate speech in the context of migration offer excellent challenges for linguistic analysis that stand in a curious dialectic with legal and criminal investigation. This relationship is another version of the one-way directionality from language to the law, as is so often the case in the vast contact field between language and law (Stein 2021). In the following sections, we will add some aspects that highlight this unidirectional interdependence of language and law.

5 Linguistic aspects of the principle of legality

It is not just that the particular crimes that are the subject of this volume are, to a large extent, committed by using language. But also the meta-fact that the terms and categories of analysis and description that actually constitute the crime (for instance, exactly when do we speak of *inciting*?) are logically and primarily established through linguistic analysis, which then forms the basis for legal and criminological classification. It is the linguist's task to demonstrate how the physical act of producing a linguistic utterance is interpreted by the utterance's recipient This already leads us to the heart of the matter: it is the linguist's job to explicate how a certain utterance (spoken, written, digital) is understood by the reader or hearer. Linguistic analysis offers a conceptual template and sets of categories that can be compared and related to the conceptual grid defined by legal and criminological research (equally established in and through language). And we always have to bear in mind the additional complication of remaining aware of whether we are talking about the native speaker's "naïve" scientifically unreflected understanding, which is actually what the linguist must take apart, or about the legal analysis with a professional eye, applying the professional dogmatic categories of research, as a Procrustean bed to establish a fit between perceived facts and pre-existing analytical categories in the individual case. As will be seen, there is a significant degree of homology, but no one-to-one correspondence. The crucial difference arguably arises from the legal principle of legality. As Doval Pais writes in Chapter 9 in this volume: "The task of determining the legal meaning (interpretation) should remain within the margins of the possible literal wording of the legal terms, as the principle of legality requires".

This raises another critical point about the relationship between legal and linguistic analysis and concepts. If we interpret Doval Pais' exposition correctly, the legal principle of legality would require that criminal law, as a law of facts, uses clearly identifiable criteria and unequivocal terms and descriptions to define crimes, such as finite lists of crimes. This is why the legal world sets great store in pointing to the literal senses and uses of terms. In other words, the literal meaning of words and sentences must be the ultimate guiding principle for legal interpretation, as anything not captured by the default extension of the term will require extraordinary interpretive efforts.

Such reasoning is, however, based on assumptions about literal meaning—about what literal meaning could even be—that have been challenged by linguists who assail its usefulness as an analytic construct (Recanati 2004; Stein 2017). Something that is a central dogmatic—and one would add, necessary—tenet in law is problematic from the point of view of linguistic analysis.

This is, therefore, one of the critical areas at the law-language interface where linguistic analysis and legal categorial practice are not in lockstep. However, in practical usage terms, both normal, naïve language users and lawyers depend on a notion of literal meaning when talking about language. For the law, it is what might be called an "operatively necessary assumption". That is, a notion of the literal is a foundational part of the dogmatic architecture, without which the law might not be able to function.[7]

6 Functional stages

While the immediate impetus to study hate crimes are physical acts of violence, overt verbal expressions of hatred, and actual language crimes related to or caused by hate,[8] this group of phenomena can be seen as the terminal point or peak and consequence of a chain of phenomena (in both logical and temporal order) leading up to this end result. Although it is beyond our competence to discuss phenomena in terms of social psychology, it seems safe to say that at the beginning of this chain of phenomena (stage 1) there are attitudes or predispositions or the existence of vulnerable groups which may foster negative attitudes, or

---

7 For a recent discussion of concepts of "literal" in legal interpretation see Smolka and Pirker (2021).
8 Compare the notion of "motive" in Chapter 9 in this volume.

aggression or hate towards migrants which may surface in negative speech acts. This hotbed of negative feelings, conceptualised as "emotions", will be manifested in speech acts expressing negative views about people perceived as minorities (who may demographically be a majority). The next step in this chain of phenomena (stage 2) may be the actual rise of hate in public speech acts through language use that can globally be termed "hate speech"—speech acts that can potentially express hatred and generate hate in larger sections of the population. This may lead to (stage 3) physical acts of violence.

Language is likely involved in several functionally distinct ways in this sequential process. It is, therefore, essential to invoke a linguistic concept that provides an analytical toolbox to help bring out the different functions of language in this process: *Speech Act Theory*. There is no need to elaborate here on this pragmatic theory of language use, as this is done in this volume by Guillén-Nieto in Chapter 8 and in many representations accessible to interested non-linguists. Briefly, in language use there is a logical and functional distinction between the locutionary act (the physical uttering of language, the grammatical content), the illocutionary act (the intended type of action of how the hearer is to understand the speech act), and the perlocutionary act (what the effective intended consequence of the speech act is). These acts are simultaneous (not temporally sequenced) and logical levels of analysis of any individual utterance of language.

Speech acts can be conceptualised as being ordered hierarchically, such that a local speech act (like a sentence) may receive its final interpretation through being part of a larger text-sized or genre-sized speech act, in the way spelt out in the contribution to this volume by Guillén-Nieto. Speech acts in stage (1) are essentially of an expressive or representative nature: they describe people or actions perceived as states of affairs or express their subjective feelings. Speech acts at stage (2), while having an expressive component (as a necessary condition), are essentially directive in nature: the definitional feature of speech acts in stage (2) is their character as incitement to either or both feelings of hate and hate-motivated physical action. Incitement and physical action may also be seen as internally sequentially ordered.

7 The linguistics of actionability

The first application of Speech Act Theory to the analysis of hate speech is in bringing out the different types of language used at the other logical stages in reactions to migration. The next application of Speech Act Theory directly addresses the issue of legal actionability. While the legal and penal side of the criminality of hate speech is directly addressed by Doval Pais in this volume, a brief discussion of the logically prior status of the language used at these three stages is in order. At this point, Speech Act Theory and the more general perspective of the pragmatics of language use (see Sperber & Wilson 2012) together offer an insightful perspective for a forensic approach to hate speech.

The contact zones of language and crime, as reflected in the area of forensic linguistics, are of a fundamentally twofold nature:

(1) Language used to perpetrate crimes.

(2) Language used as evidence to investigate crime as part of police or court activity.

Both aspects are relevant in the types of cases discussed here. In (1), two subtypes must be minimally distinguished:

(1a) Uttering language (the locutionary act) is a crime, such as when individual expressions are legally banned. The prohibited act is committed by physically realising the phonetic form in spoken language or in graphemes representing the concept in written or Internet-scripted form.
(1b) All utterances or discourses with legally defined content are prohibited (for example, Holocaust denial).

Normally, these legal prohibitions are culture-specific, i.e. relative to individual countries and to historical and political situations. They are variously enshrined in formal statutes as prohibitions and are penally sanctioned. This category (1a and 1b) includes negative speech acts like defamation, hate speech and insult, but also involves cases in music, literature or other arts.

A main difference between types (1a) and (1b) resides in that (1a) is more literal, i.e. physically visible and manifest. In contrast, type (1b) as a rule, requires interpretation, sometimes on a major scale, such that there is no immediate or direct link between the words on the linguistic surface and the function (e.g. directive) of the utterance, be it an individual utterance or a discourse in its entirety.

This difference in directness and immediate meaning or functional accessibility makes a significant difference for identifying and resolving crime (2). As with any other crime, forensic linguistics has the task of identifying traces of a crime, which the court may award the legal-procedural status of evidence (as is done by a jury in the Common Law system and by judges in the Civil Law system).

While the job for the forensic linguist is easy in type (1a) language crime, it is a major definitional challenge for the legal system in the type of language crime at the heart of discussion in this volume, type (1b). It is the task of the forensic linguist to identify arguments as to why a specific utterance should have the character of "hate speech" or why it should have "the potential for the act to be carried out" (see Chapter 9), and even more precisely how much of this potential is contained in the words themselves (i.e. in the locutionary act). This task is not normally seen as one for the forensic linguist, but it is up to the linguist to supply theory-based tools of analysis that are suited to provide the court with arguments (ultimately linguistic "traces") for classing a discourse as an instance of hate speech. This certainly widens the brief of the forensic linguist. Still, the court is looking for arguments for the status of an utterance as a hate speech crime or one that has the potential to lead to such a crime. In that case, it clearly is a linguistic-forensic activity to identify properties of the utterance itself and its meaning-determining relevant contextual factors that may be relevant for a decision whether or not to class the utterance as hate speech.

8 Indirect vs. literal

Of particular relevance—and arguably the hardest and most critical part in analysing type (1b)—is contained in functional stages (2) and (3) of a speech act. Stage (1) is characterised by a large class of negative speech acts that—however—do not normally have legal relevance: speech acts expressing prejudice, dislike, or even hate in the form of expressive and representative assertion, such as "I hate immigrants" or "immigrants are only up to exploiting our social systems". These speech acts may be directed at any other group that may be socially stigmatised. The exact nature and kinds of expression are mostly of interest to the linguist, but are not normally interesting from the point of view of legal actionability; they share an important structural problem with language use in stages (2) and (3)—namely their indirectness. While type (1a) is mostly perpetrated by uttering the incriminated form, and the content is then directly evoked—the path from the lexical meaning of the word to comprehension being very short—with type (1b) utterances, it is uncommon to commit the speech act in a direct form by uttering the word, phrase or sentence that conventionally designates the incriminated content.

Thus, for instance, few people would say, "I threaten you with xy". The standard way of threatening (Muschalik 2018) is to not use expressions that have as part of their lexical semantics an element that alludes to the function—and in this case the directive function—of the speech act, such as "I know where your daughter lives". Speech acts can be performed with various degrees of directness all the way to not at all literal expression of the illocution. And it is a hallmark of linguistic expressions in this area that the speech acts are expressed with a very high degree of indirectness. The analysis of hate speech is one of the most complex tasks faced by the linguist and pragmatician, because the task is the reconstruction, from the locutionary act (what has been *said*) in conjunction with all aspects of language-external context assumptions, of the ultimate intention as represented in the illocution.

To be more specific, the linguist's job is to make explicit—to lay open to the linguistically trained analytical eye—the inferential pathway that a presupposed normal default person (not necessarily a lawyer) would follow in constructing the final intended utterance meaning. Thus the hearer starts from "I know where your daughter lives" as the locutionary act to arrive at the intended illocutionary act "I herewith threaten to abduct your daughter if you don't pay me a certain sum of money". This, then, is the final inferred utterance meaning. Note that the normal language user is not in a position to be aware of the complexity of the inferential process in all its hypothesised steps and information flows, as is made explicit in Guillén-Nieto's Figures 8.2 and 8.3 (in Chapter 8), which represent the hierarchical structure of the construction of the final utterance meaning. The hearer's work is to rebuild, in reverse, the path of generating hate speech. There is a double sequentiality involving two different analytic dimensions. The first dimension of sequentiality is a social or actional one, extending from the fostering factors or vulnerabilities to the rise and manifestations of hate feelings, designated above as stage (1), to verbal actions as stage (2) and ultimately to stage (3) phenomena of violent action.

To the extent that verbal behaviour is involved, there is also a linguistic sequentiality that the normal language user engages in naively, and which the linguistic expert must explicate in its derivational or inferential steps. The linguist's job here is to bring their expertise in retracing inferential steps, as they were ultimately triggered by the semantic potential of the expressions used in the locutionary act and realised in the ensuing inferential steps. In particular, *linguistic* refers to all knowledge and information flowing from the grammatical to the cognitively present other types of knowledge. Hearers are not normally aware of their own complex knowledge management, as these processes are subconscious and not accessible to the average speaker. In addition, the sequentiality involved is not necessarily real-time but primarily logical and informational. The task of the forensic linguist is therefore not easy: it is their job to identify elements of the locutionary act (the form or surface of language, the locus of literal meaning) that can form a trace of the crime of hate speech.

It is not enough to point to static elements of contextual knowledge that could plausibly have fostered an interpretation of hate speech (that can be an element in legal interpretation). Still, for a linguistic crime to be prosecuted, it must be possible to identify physically produced language with a specific hate speech crime. The question put to the forensic linguist is therefore: How can elements of the locutionary act be standardly or conventionally associated with a hate crime?

In order for forensic analysis to arrive at a high probability that a given linguistic structure is a likely or certain index of a hate crime, it would be necessary to establish regularities linked to specific types of contexts, presupposing actual experience. This would amount to fulfilling the postulate formulated by Doval Pais (Chapter 9) that there should be regularities and elements of literalness in committing the crime, which enhance the probability of a correct interpretation of the author's intention. These regularities would be of two basic types. The *locutionary* type, where the semantics of individual expressions would yield an indication of a potential misdeed, such as, for instance, the mention of "your daughter" or "a sum of money". The other, the more indirect type, will not necessarily contain such lexical material but would be established by recurrent instances of using specific linguistic material in established hate speech cases in a given historical context.

However, in most cases, given the diversity and the range of differences between individual circumstances, one has to concur in principle with Doval Pais (Chapter 9) that each case must be followed through explicitly, in what linguistic pragmaticians call the *meaning-making* process, making the expertise of a scientist trained in the pragmatic analysis of utterance meanings and the inferential processes involved even more essential. This will ultimately result in a typification that allows for the setting up of class types in a territory that has been largely neglected by linguists, who have been mostly concerned with analysing positive speech acts not negative ones.

9 Intention

The term hate speech, as it has been used up to this point in the discussion, needs to be specified and differentiated linguistically as to its internal structure, and as to the appropriate differences in legal consequences. While what was termed stage (1) is mostly a matter for the sociologist or social psychologist, stage (2) is characterised by actional or dynamic features of language use. In terms of speech acts, stage (2) is predominantly characterised by speech acts with directional illocutionary character—speech acts that aim to induce in other speakers changes of attitudes, changes which may inspire physical acts of violence (stage 3). Guillén-Nieto makes clear in her contribution to this volume (Chapter 8) that this is where the greatest challenges lie for linguists and lawyers. What matters centrally for both professions is that the speech acts in this logical phase are intended to incite hatred and, ultimately, violence.

In that respect, an analogy to the act of lying may be helpful here. Whether an attempt to lie is successful is not a matter for consideration from the linguist's point of view, but arguably from the moral or legal point of view. The litmus test is the analogy to the case of accidental truth discussed by Horn (2017). Is an assertion a lie (linguistically? morally? legally?) if it was intended as a lie, but is factually (but not in the mind of the speaker) the truth? Is objective factuality the criterion or is it intentionality?

When it comes to the properties of the locutionary act in the case of hate speech, from the point of view of the law, they must be purposed to incite hatred. For the linguist, it is not essential that they actually incite hatred but that they have the potential to incite hatred. In terms of Speech Act Theory, what matters is the illocution: that it be intended as a directive speech act. The speakers use expressions in such a way that these words, through their grammatical or semantic potential (as manifested in a history of use of such terms) are seriously intended to (at least) lead to a specific change of mind or attitude. The actual success of the intention is the performative act (to be performed by the hearer/reader). What matters for the linguist (perhaps also for the lawyer) is that the hearer must reconstruct, based on the utterance and the context, that (for the lawyer: "if") they presume the speaker (writer) to have chosen linguistic expressions such that they are intended to induce in the hearer feelings of hate towards a certain group, such as present in stage (1). The essential point is not even whether or not they could be intended, but whether they are intended in a specific instance of usage. What the forensic linguist seeks are traces of this inferential process adequate to be counted as evidence.

If this is answered in the affirmative, then we have hate speech. It does not matter whether or not hate is generated and the performative act on the speaker's side is completed. For the pragmatics-oriented linguist, the presumptive, reconstructed intentionality is the touchstone. Hate speech, as incitement to hatred, is therefore located in the illocutionary force attributed by the speaker—and by the analyst—to the utterance or to the whole discourse. So, from the linguist's point of view, actionability or guilt is incurred by the specific illocutionary character of the speech act. It defines the *mens rea*. So, a speech act is not inherently an incitement to hatred but is *intended to be interpreted as* an incitement to hatred. Bringing forward the traces of this process clearly, in such a way that they can be reliably counted as evidence, imposes a difficult task on the linguists and their forensic analysis.

While this interpretation is reasonable from the linguist's point of view, it may not be quite clear how the lawyer could follow along, given the tendency in law to prosecute concrete completed actions and to locate meaning in what they consider literalness. There does seem to be a minimal *ad hoc* consensus that there cannot be a purely platonic, a-contextual, purely compositional level of meaning as the locus of literal meaning. While most pragmatically oriented linguists would argue that literalness in the sense of completely a-contextual, purely language-based (i.e. compositional) level is an ineffectual analytic construct—impossible to pin down in rigorous terms—lawyers (and normal language users) would find a life without some practical and legally manageable notion of literalness unmanageable.

What puts the ball back into the linguist's court is the legal requirement that the speech act should have the potential to be interpreted as incitement to hate. It is appropriate here to cite a famous German case from 1970s, when the political climate in Germany was especially hot. The court recognised the hate speech potential of a 1967 leaflet that asked why a certain Berlin department store was not yet on fire (*"Wann brennen die Berliner Kaufhäuser?"*). The leaflet was distributed in the student union buildings of Berlin universities. The court made it clear that it assumed an interpretive community of normal citizens who clearly would have construed the potential directive character as a call for arson. However, the court felt unable to arrive at a guilty verdict, as it found it impossible to find clear and concrete evidence of the intention of the leaflet's authors to direct people to commit arson. Although the call for arson was arguably as close "in the language" as it could be, the court felt there was no evidence for the intention—a state of mind—to call for physical action and voted for the primacy of freedom of expression. So, although there was both locutionary evidence (*"Kaufhäuser"*, *"Brennen"*), and there was at this historical point in time a tradition (a regularity) of discourse on setting department stores on fire (plus other non-linguistic crimes), the court decided they had no evidence of the intention to generate hate and incite physical criminal action. One wonders if today's legislative and historical situations would have provided a backdrop for a different decision.

In the case from Germany just cited, there was as much literal evidence as possible as an indication of the potentiality of the verbal utterance. But the court did not find that there was evidence to assume intentionality to incite hate or physical criminal action. This appears to be a crucial issue for the law, as there is a discussion about whether they are discoverable (or known, as descriptivist or cognitivist theses argue) or imputable (as ascriptivist or non-cognitivist theses argue) (cf. section 2, above). Discoverability therefore includes not only locutionary aspects but also language-external knowledge that the lawyer and linguist suppose was present at the moment of uttering, such as a responsive audience as a social hotbed.

However, like threats, incitation to hatred is, of course, a social act and arises only in a specific social context. This is why Bonsignore's distinction (Chapter 2) between an abstract and a social threat is pertinent: only threats and inflammatory speech in a given social context can be intended as incitation, and only then can they linguistically and legally count as such. This is why testing utterances in abstract conditions, or considering their semantic potential *in abstracto*, cannot provide actual evidence. This puts the forensic linguist in a difficult position: the task is to find traces in the actual choice of expressions that were likely to be interpreted as evidence of the intention of the speaker to incite hatred in a given—and concrete—group of hearers. This presupposes that the analyst has expert knowledge in a typology of which expressions can be used with a specific group of language users. This fact points to a further dimension: the retroactive construction of an intended illocution is always relative to a certain assumed audience. Bonsignore's notion of a "social threat" (Chapter 2) always implies a presumed audience.

10 This volume

This introduction has attempted to weave together several issues taken up in more detail in the individual papers of this collection, into a wider view of the phenomenon of hate speech from three major perspectives: language, discourse, and law. A central role is, of course, accorded to the conceptualisation of issues and players as reflected in the semantics and use of individual terms on both the object level and the meta-level. To start the volume (Chapter 2), Bonsignore aptly gives us some historical-semantic underpinning. He shows how the etymology of the word "danger" has a two-fold developmental lineage, providing a conceptual pedigree for an important legal distinction significant in the discussion of "danger": abstract danger and danger in the sense of actual, real-time threat as perceived in a social context. The distinction has significant legal consequences.

The subject of the discussion in the present volume, and the frequent trigger and object of hate speech, is, of course, dependent on the recognition and physical presence of migrants. Schmid's paper (Chapter 3) addresses a central practical issue: how do we know that a person, often enough with no documentation from their country of origin, is a legally qualifying migrant? Again, language plays a central role. Schmid demonstrates convincingly—and worryingly—that traditional non-expert views about the availability of the language of origin, such as are widely used in assessments of the country of origin, can be quite misguided in assuming the persistence of the individual speaker of the language of the country of origin: the fact that there is no more competence in the presumed language of provenance cannot automatically be taken as negative evidence that a person does not or cannot have come from a certain country. Schmid presents ample empirical evidence that non-expert assumptions can be quite misguided, with dismal effects on decisions about individual persons—one of many misconceptions about the nature and function of what "language" is and its role in the ambit of migration and hate speech.

Hate speech as incitation to hatred does not come out of nowhere, but grows in societal hotbeds, here designated as factors in stage (1). Two papers address the manifestations of this hotbed. Hartwell (Chapter 4) examines the use of referential terms by lawyers and supreme court judges. While one would want to concur with her statement that further study is needed to analyse the use of grammatical items like pronouns, she does provide adequate evidence to demonstrate that the linguistic manifestations of conservative and progressive ideologies are manifested in the use of the term "aliens" on the one hand and terms like "non-citizens" on the other.

Lavissière and Hill (Chapter 5) provide access into an even deeper layer of the hotbed, finding a hotbed for hate speech on the level of the creation of laws. The authors deal with what Doval Pais calls "the reasons behind the existence of these criminal rules". Uncovering the deepest propitiating level of the process described as stage (1), Lavissière and Hill identify the use of terms in the semantic area of immigration that is characterised by a "racist animus" linking "immigrant" to racially discriminatory language, which arguably threatens the very validity of these laws in today's legal climate.

The relative scarcity of the linguistic material and the high amount of inferentiality—with the incumbent difficulties for linguistic forensics—are in fact a hallmark of incitation to racial hatred. As is shown by the contribution by Ruzaite (Chapter 6), who applies a typification developed in Impoliteness Theory to a corpus of hate speech directed at migrants from Lithuanian news portals, there is at this time no way to establish conventionalised formulae of hate speech such as could be utilised to automatically link language use to criminal acts. Such a system would be of great value in establishing reliable types of evidence for use in criminal proceedings. As it stands, every case needs to be analysed on its own merits, and the inferential process has to be made explicit in each one, something which constitutes a major challenge to the analytic competence of the linguistic-pragmatic analyst. The challenges are even exacerbated by the fact that perpetrators may, even tactically, engage in deliberate efforts not to offer points of attack for actionability. It remains to be seen what further challenges or opportunities are provided by the KI facilities: Can chatbots be trained and taught to produce hate speech?

The problem of the camouflaging of the intent to incite (the specific illocutionary force), which so often characterises hate speech and constitutes the core problem in attempts to tie hate speech to conventionalised linguistic surface forms, is also at the centre of Baider's (Chapter 7) analysis of Islamophobic and anti-Semitic discourses. She discusses the very nature of what she terms "covert" hate speech and argues for using conspiracy theory as the vehicle for analysing hate speech, much as Ruzaite employs (Im)politeness Theory.

Finally, the last two chapters define the theoretical constituent core of the phenomenon from the two major perspectives of the volume, as will have become clear from the preceding sections of this introduction. Guillén-Nieto (Chapter 8) analyses the linguistic and pragmatic side of the phenomenon of hate speech.

Her question: exactly how do we get to interpret and award a discourse the quality of hate speech and thereby identify it as a language crime? Doval Pais (Chapter 9) disentangles the complexities of the legal and criminological side of hate speech, demonstrating why and how this phenomenon of language use is actionable. Both papers highlight in particular the importance of precise and carefully considered use of language in drafting applicable laws and conducting the legal process in dealing with hate speech.

It is hoped that this volume will be a step toward a linguistically and legally informed treatment of a painful problem in today's society.

References

Alba-Juez, Laura & Geoff Thompson. 2014. The many faces and phases of evaluation. In Geoff Thompson & Laura Alba-Juez (eds.), *Evaluation in context*, 3–24. Amsterdam: John Benjamins.

Alba-Juez, Laura & Tatiana Larina. 2018. *Language and emotion: Discourse-pragmatic perspectives*. 22(1). 9–37.

Arzoumanov, Anna. 2021. Freedom of art in French legal proceedings: A discourse analysis perspective. In Janet Giltrow, Frances Olsen & Donato Mancini (eds.), *Legal meanings: The making and use of meaning in legal reasoning*, 17–30. Berlin/Boston: De Gruyter Mouton.

Assimakopoulos, Stavros, Fabienne H. Baider & Sharon Millar (eds.). 2017. *Online hate speech in the European Union: A discourse analytic perspective*. Cham, Switzerland: Springer.

Balibar, Étienne. 2006. Strangers as enemies: Further reflections on the aporias of Transnational citizenship. *Globalization Working Papers* 6(4). 1–17.

Barata, Joan. 2020. Informe jurídico sobre la adecuación a los estándares internacionales en materia de libertad de expresión de determinados preceptos del Código Penal español. Plataforma en defensa de la libertad de información. https://libertadinformacion.cc/wp-content/uploads/2020/02/Informe_PDLI_C%C3%B3digo-Penal_JBARATA_2020.pdf (accessed 15 January 2023).

Brandenburg v. Ohio (1969). Supreme Court of the United States. https://supreme.justia.com/cases (accessed 21 January 2023).

Cannon, Walter B. [1915] 1925. *Bodily changes in pain, hunger, fear and rage*. New York: D. Appleton & Company.

Delgado, Richard & Jean Stefancic. 2004. *Understanding words that wound*. New York: Routledge.

Delgado, Richard & Jean Stefancic. 2018. *Must we defend Nazis? Why the First Amendment should not protect hate speech and white supremacy*. New York: New York University Press.

Erbakan v. Turkey (2006). European Court of Human Rights. Application No. 59405/00. https://hudoc.echr.coe.int/app/conversion/pdf/?library=ECHR&id=003-1728198-1812055&filename=003-1728198-1812055.pdf (accessed 15 January 2023).

Ferrajoli, Luigi. 2019. Políticas contra los migrantes y crisis de la civilidad jurídica. *Revista Crítica Penal y Poder* 18. 182–193.

González Lagier, Daniel. 2004. La prueba de la intención y el principio de racionalidad mínima. *Jueces para la democracia* 50. 41–51.

Guillén-Nieto, Victoria. 2023. *Hate speech: Linguistic perspectives*. Berlin/Boston: De Gruyter Mouton.

Hess, Ursula & Shlomo Hareli (eds.). 2019. *The social nature of emotion expression*. Cham, Switzerland: Springer.
Horn, Larry. 2017. Telling it slant: Towards a taxonomy of deception. In Janet Giltrow & Dieter Stein (eds.), *The pragmatic turn in law: Inference and interpretation in legal discourse*, 23–46. Berlin/Boston: De Gruyter Mouton.
Ibero-American Court of Human Rights. Advisory opinion OC-05/85. https://www.corteidh.or.cr. (accessed 15 January 2023).
International Convention on the Elimination of All Forms of Racial Discrimination. 1965. https://www.ohchr.org (accessed 20 January 2023).
International Covenant on Civil and Political Rights. 1966. https://www.ohchr.org (accessed 20 January 2023).
Pirker, Benedikt & Jennifer Smolka. 2018. International law, pragmatics and the distinction between conceptual and procedural meaning. *International Journal of Language and Law* 7. 117–141.
Kopytowska, Monika & Paul Chilton. 2018. "Rivers of blood": Migration, fear and threat construction. *Lodz Papers in Pragmatics* 14(1). 133–161.
Mackenzie, J. Lachlan & Laura Alba-Juez (eds.). 2019. *Emotion in discourse*. Amsterdam: John Benjamins.
Matsuda, Mari J. 1989. Public response to racist speech: Considering the victim's story. *Michigan Law Review* 87(8). 2320–2381.
Matsuda, Mari J. [1993] 2018. Public response to racist speech: Considering the victim's story. In Matsuda et al. (eds.), *Words that wound: Critical race theory, assaultive speech, and the First Amendment*, 17–51. London: Routledge.
Monnier, Angeliki, Axel Boursier & Annabelle Seoane (eds.). 2022a. *Cyberhate in the context of migrations*. Cham, Switzerland: Palgrave Macmillan/Springer.
Monnier, Angeliki, Axel Boursier & Annabelle Seoane (eds.). 2022b. Anti-migrant hate speech as a symptom of a representation crisis. In Angeliki Monnier, Axel Boursier & Annabelle Seoane (eds.), *Cyberhate in the context of migrations*, 1–17. Cham, Switzerland: Palgrave Macmillan/Springer.
Muschalik, Julia. 2018. *Threatening in English: A mixed-method approach*. Amsterdam: John Benjamins.
Recanati, Francois. 2004. *Literal meaning*. Cambridge: Cambridge University Press.
Schröter, Melani. 2022. "Everything goes against the German here!": Self-victimising discourse in comments on migration-related posts on the alternative für Deutschland Facebook page. In Angeliki Monnier, Axel Boursier and Annabelle Seoane (eds.), *Cyberhate in the context of migrations*, 83–114. Cham, Switzerland: Palgrave Macmillan/Springer.
Smolka, Jennifer & Benedikt Pirker. 2021. Pragmatics and the interpretation of international law: Two relevance theory-based approaches. In Janet Giltrow, Frances Olsen & Donato Mancini (eds.), *Legal meanings: The making and use of meaning in legal reasoning*, 131–163. Berlin/Boston: De Gruyter.
Wilson, Deirdre & Dan Sperber. 2012. *Meaning and relevance*. Cambridge: Cambridge University Press.
Steiker, Carol. 1999. Punishing hateful motives: Old wine in a new bottle revives calls for prohibition. *Michigan Law Review* 97(6). 1857–1873.
Stein, Dieter. 2017. On inferencing in law. In Stein Dieter & Janet Giltrow (eds.), *The pragmatic turn in law: Inference and interpretation in legal discourse*, 335–368. Berlin/Boston: De Gruyter Mouton.
Stein, Dieter. 2021. Sprache und Recht: Das Recht als Forschungsobjekt der Sprachwissenschaft. In Daniel Greineder, Karl Pörnbacher & Stefan Vogenauer (eds.), *Schiedsgerichtsbarkeit und Rechtssprache: Festschrift für Volker Triebel zum 80 Geburtstag*, 207–318. München: Beck.
White, Peter R. R. 2002. Appraisal: The language of evaluation and stance. In J. Verschueren, J.-O. Östman, J. Blommaert & C. Bulcaen (eds.), *The handbook of pragmatics*, 1–23. Amsterdam: John Benjamins.
Wodak, Ruth. [2015] 2020. *The politics of fear: The shameless normalization of far-right discourse* (2nd ed.). London: Sage.

Dyango Bonsignore
Chapter 2
The dangerous individual in a time of migration: *Periculum, dominium* and dangerousness

1 Introduction

This chapter analyses the relationship between the etymology of *danger*, the theoretical reflections regarding the historical uses of this notion, and the applicability of all this to the current migration crisis. More precisely, I first set the conceptual stage by providing a brief examination of two main etymological branches that correspond to the idea of danger: *periculum* and *dominium*. These are presented for the purpose of extracting some theoretical insights from the historical uses of language to convey and designate danger. These insights are then applied in the following section to draw a connection to historical and contemporary debates in the field of criminology. The idea of (criminal) dangerousness serves as the link to connect the uses of language, the social process of naming and labelling, and the criminological analysis of what could be called *dangerousness discourse*. Finally, the underlined insights are applied in a tentative reading of the migration crisis, through the lens of what we will call *dangerisation* and *endangerment*. Conceptual connections to *periculum* and *dominium* are used as anchor points to analyse some of the characteristic features through which the migration crisis has manifested in recent times.

In the social sciences it is not unusual to confer a certain degree of importance to words and to the social uses of words. This is unsurprising, considering that words are sometimes all we have with which to construct reality. One need not subscribe to a Baudrillardian view that portrays social reality as mere "simulacra and simulation" (Baudrillard 1981) to recognise the importance of language in

---

Note: This paper arises from the research projects: "Manifestaciones de desigualdad en el actual Sistema de justicia penal: examen crítico de las razones de necesidad, oportunidad y peligrosidad para la diferencia (AEQUALITAS)" (ref. RTI2018-096398-B-I00) founded by the Spanish Ministry of Science and Innovation, and "Víctimas especialmente vulnerables e indicadores para su determinación" (ref. GV/2021/133) founded by the Valencian Office of Innovation, Universities, Science and Digital Society.

Dyango Bonsignore, University of Alicante

https://doi.org/10.1515/9783110789157-002

shaping human understanding and behaviour. Usually, even in highly technological and scientifically sophisticated societies, we invest a fair amount of trust in language, being the most immediate way to deal with day-to-day events. In such a broad and general sense, it is hard to argue that, as far as human interpretation is concerned, reality is in many important ways a social, linguistic construction (Berger & Luckmann 1967).

This knowledge has two main ramifications for social-scientific thought. The main one relates to its principal object: because humans interact mainly using language, linguistic objects in themselves become points of inquiry. In this sense, we might wonder about the ways in which people *name* certain objects, how this has changed over time, how many concurrent names (and definitions) can be related to a single object, or how all this shapes the perspective of the speaker. Briefly, at issue here is the social construction of reality itself, to the extent that it is made by people through language.

The second ramification is of an epistemological nature, making *social science* an object of inquiry *for* the social scientist. In this case, language is still a point of inquiry, but it is the more systematic language found in theoretical and empirical discussion. Given that raw *data* require interpretation, and such interpretation is usually conveyed through language, scientific communication and verbalisation is just as much a legitimate object of analysis as "folk" communication is. Struggles to establish definitions, discussions of the limits and uses for scientific concepts, formulating hypotheses, drawing conclusions—these are all examples of uses of language that can be fruitfully studied without venturing into the domain of postmodern linguistic and social theory.

One could argue that the difference between studying "folk" communication and scientific discourse is not one of quality but of perspective; in the first instance, the gaze of the social scientist is directed outwards (to society at large), in the latter case it is directed inwards (to the social scientist and its particular domain of social interaction). Moreover, in cases where the scientific and the "lay" use of words and concepts are substantially the same, such distinctions may be merely academic.

With all this as a starting point, there is a large and well-established tradition of philosophical and social research into the history of scientific concepts. In this sense, it is not unusual to find in the history and etymology of words starting points for productive theoretical inquiry. Some schools of thought have taken this much further, building entire works on unearthing the historical processes that have intervened in the formation (or transformation) of a concept.

In criminology, the work of French philosopher Michel Foucault has been foundational. The keyword for describing Foucauldian inquiry has been *genealogy* (Foucault 2004: 22–4; Koopman 2013: 18). From this point of view, one would hope to find, in studying the history of a concept, the roots of its present form and, possibly, some forgotten purpose—an inertia from the past that continues to shape its current uses. To put it in a somewhat literary fashion, these inertias and forgotten purposes hidden in words can be like the "unconscious" of common sense.

But one could say that the purpose of unearthing the deeper motives behind present conditions through retrospective research (be it strictly historical, or more akin to the history of language) may be shared without the need to subscribe to a framework such as the aforementioned. There is room for contributing to this general direction of inquiry without tackling every conceivable dimension. Globally, theoretical research can draw useful information from delving into the actual uses of language.

This chapter will try to advance along the middle path between a very specific social-scientific standpoint (that of criminology) and the search for theoretical clues drawn from the history and etymology of words. More to the point, this chapter will try to deal with the recently reinvigorated notion of "dangerousness", stitching together an inquiry into past uses of the word *danger* with a brief account of the criminological tradition. Nonetheless, as we have already suggested, an underlying worry about present social conditions, exemplified in the migration crisis, fuels this research. Not surprisingly, then, we will try to bring the aforementioned analysis as close as possible to what constitutes one of the main manifestations of contemporary dangerousness discourse: human mobility, and what has often been called, compellingly, *crimmigration* (Stumpf 2006: 376).

2 The etymological clue: *periculum* and *dominium*

From a criminological standpoint, the discussion in question is not without precedent. As a matter of fact, interesting and fruitful insights into the etymology of dangerousness were first pointed out by Theodore Sarbin (1967), and again a decade later by Ysabel Rennie (1978). That being said, their efforts were barely recognised within the field, and failed to inspire much criminological reflection. Here I would like to remedy that situation, and pick up from where they left off. To that end, I begin by examining (in a succinct way, from the perspective of the criminologist, not the philologist) two distinct etymological pathways: one related to the English and French words *danger*, and the other to the Spanish *peligro* or the Italian *pericolo*.

2.1 Periculum

Starting with the latter two, *peligro* is the characteristic word to express danger in Spanish. *Peligro* can also be used to convey other notions that lie in the semantic vicinity, such as: risk, contingency, threat, or insecurity. Accordingly, the word is defined in the Royal Academy's Dictionary of Spanish Language as the imminent risk of an evil or harmful event, or the place, object or situation that increases the possibility of imminent harm. Thus, *peligro* is as much the *thing* that endangers as the *idea* of danger itself. All this may already be quite familiar and recognisable to an Anglophone.

Etymologically, *peligro* can be traced back to the XIII century's *periglo*[1] and, ultimately, to the Latin *periculum*, meaning 'trial', 'test' or 'experiment', also in connection with *peritus*, meaning 'expert' or 'experienced' (Corominas 1987: 448; de Vaan 2008: 446). Additionally, the Latin root *per-* seems to have conveyed something akin to 'going through', 'advancing', 'trying' or 'daring', broadly expressing a forward-facing attitude, movement, or dynamism. Taken as a whole, we find in the etymology of *peligro* what the contemporary eye could see as a somewhat unexpected emphasis on the *positive* act of risk-taking. It points to the risks surrounding adventurous or experimental action, such as the perils of boldly advancing into the unknown and the experience to be gained in the process.

For the purpose of discussion, we might even take the point slightly further, drawing out two main implications. On the one hand, it could be argued that *periculum* presupposes a type of agent who is portrayed as a risk-taker, insofar as the hazards are linked to the actions of trying, daring, or going through with something. On the other hand, danger itself is represented as some sort of objective entity, like a force of nature or fate, an adversity or obstacle that is "out there" and could result in unwanted consequences. For the purposes of this chapter, this identification of objects and events as potential threats should be kept in mind.

2.2 Dominium

I will now turn to the other etymological branch suggested above and, in so doing, to the works of Sarbin and Rennie. I will begin by examining the ideas of Sarbin. In his paper, "The dangerous individual: An outcome of social identity transformations," Sarbin provides a definition of danger akin to the general

---

[1] Here the reader might recognise the similarity with the English *peril* and the French *péril* (Partridge 1966: 2366).

expectation: "We might expect the origins of the word 'danger' to be related to physicalist conceptions because of its widespread use for denoting physical objects and events that may damage property or maim and kill people" (Sarbin 1967: 286). To some extent, this would be in line with the previous examination of the etymology of *periculum*. But things are in fact more interesting and, according to Sarbin, the origins of the English and French *danger* can be traced back to the Latin *dominium*. More precisely, the English word seems to come from the Old French, which we can trace to the form *dangier/dongier* which, in turn, is connected to the aforementioned Latin expression (Partridge 1966: 724; Sarbin 1967: 286).

Although the etymological ramifications of *dominium* are complex (de Vaan 2008: 177), the line that crosses through *dongier* refers to "a lord's jurisdiction or power", coherent with *domniārium* as 'power' (mainly, of a lord) and, in turn, *dominus* as 'master' or 'lord'. To this, we could add deeper insight through Partridge's commentary on the origins of "dangerous" as someone "haughty (like a master), hence difficult, dangerous" (Partridge 1966: 724). Hoad reflects the same general idea, adding danger as "liability to punishment" (later reconfigured as liability to injury) and dangerous as "difficult to deal with or please" (later used as "reluctant to comply") (Hoad 2000: 112). As Sarbin aptly puts it, this conveys a conception of danger as "a symbol denoting relative power in a social organisation" (Sarbin 1967: 286) and emphasises the power to hurt or harm that lies in the hands of a lord or sovereign. Phrases such as "to be in someone's danger" are particularly expressive of these implied power relations, as equivalent to being at someone's mercy, or within his power, and similar to being in his debt.

As anticipated, one finds here an unexpected opening onto the kind of concepts that speak most immediately to the social scientist, as this etymological route points directly to the dangers engrained in the social structure and in those institutions that embody the unequal distribution of power. To be "in a person's danger" is to be vulnerable to the caprices of their will, to the possibility of being arbitrarily hurt or harmed as a result. Within this framework, there is nothing more dangerous than another's harmful intentions in a context where such intentions could easily materialise. Danger may lie within power, and it might even be the meaning of power itself. Power may be the possibility to inflict harm when one so desires, without which power is not true power.[2]

---

[2] The kind of power that could only apply to doing good deeds appears more like an obligation, even a burden, where one can only exert influence on others if that influence is subjected to a legitimate cause or virtuous reason. The power of the sword, proverbially, is so *because it can cut both ways*.

Of course, it could be argued that this exposition paints a kind of medieval scenario, with masters and slaves, lords and serfs, that cannot be easily transposed to social relations in modern times. Taken at face value, this objection may be acceptable, in the strictest of historical interpretations. But it is not difficult to see that there is ample room for extrapolation, insofar as what is really at stake is power imbalance and, more broadly, the system of rights and obligations tied to the social roles and identities (Sarbin 1967: 286). Here I would like to focus less on this second idea and insist on what could be characterised as *the danger of power*.

In turn, this discussion has several implications. On the one hand, danger is characterised here as the effect of another's will or actions, which are particularly threatening because they function in a context of domination or of disproportionate force. On the other hand, one could suggest that this definition of danger is conceived from the point of view of the potential risk-sufferer (that is, the victim), who is detached from the production of the risk but not from its outcomes. In this sense, danger as *dominium* requires at least (1) someone powerful enough to cause harm, (2) someone in a position to be harmed, and (3) a social context within which this kind of relationship is possible.

2.3 The ambiguities of danger

In confronting the conceptions of danger that come from *periculum* and *dominium*, we are faced with two possible interpretations that, although sharing the same conceptual basis, have different meanings. What is common to both is the idea of danger as mere risk. That is to say, danger as the uncertainty that comes from the possibility that things may go wrong. Such risk may be understood in the experimental sense of *periculum*, but is also at the heart of the kind of vulnerability that *dominium* refers to. When a person's fate depends on the power of another or others, uncertainty becomes an existential condition of the powerless.

Nevertheless, from this etymological starting point we can already see semantic diversity around the idea of danger. Sometimes, such diversity may even produce divergent views on how danger should be understood. For instance, (1) a physicalist versus a relationalist perspective on what danger is, (2) a risk-taking versus risk-avoiding point of view, (3) an agent-centred versus victim-centred perspective, and so on. In the next section I will argue that the multiple ways one could think about danger are clearly visible throughout the history of criminological thought and remain a fundamental point of contention. Broadly speaking, the definition of criminal danger, the hierarchies of the dangerous and the assessment of threats are largely consensual, legally based, and prone to moral panics (Cohen 2015). But, despite this general lack of critical attention, in the next section

I will try to paint a more complex picture of the ways the different visions of criminal danger interlock. Using historical and theoretical examples, I will try to underline the key ideas that will be useful in the last section.

3 Criminologies of danger

When one turns to criminology to better understand the dominant representations of criminal danger, the first reasonable doubt that arises is one of definition and delimitation. In particular, it can be argued that, in a broad sense, the discipline as a whole is always concerned with criminal danger. Perhaps by definition, crime and criminals tend to be considered from the point of view of danger. Most criminological theory, in turn, can be seen as a way of analysing criminal danger, explaining its causes, and assessing its foreseeable consequences. Here the literature is beyond reach, and the perspectives are of all shapes and styles: from the explanation of youth crime from a developmental perspective, to gang violence and urban ecology, mob violence, cybercrime, etc. From this, one could surmise that almost every phenomenon that comes under criminological scrutiny is aimed at assessing and managing risks.

That being said, a more restrictive approach to danger within the field of criminology might be more helpful, although it could also be more problematic. From this strict perspective, criminology has often been concerned with "dangerous offenders", implying that dangerousness should be seen as an adjective, a qualification that characterises only a specific subset of offenders. The underlying assumption is that there exists something like an ordinary criminal who should not cause worry, and also a dangerous one who requires strict supervision and control.

3.1 The standard theory of criminal dangerousness

Although the following commentary does not reflect an accepted terminology, I would argue that, on the topic of criminal dangerousness, there is a "standard" theory that functions as the underlying narrative of the current consensus. Specifically, in criminal law and criminological research, the dangerous offender topic is concerned with two types of offenders. On the one hand, the *serious sexual and violent offender*. On the other hand, the *terrorist*. I would even argue that, although the danger of the terrorist is axiomatic in the field (more so in legal and political commentary, as well as in security studies), it is mostly the sexual or violent offender who is approached explicitly through the framework of dangerousness. Some recent publications that have "dangerous/dangerousness" as their main subject can be cited as evidence on this point (Campbell & Messing 2017; Harrison 2011; McSherry & Keyzer 2011; Van Brunt 2015).

It is certainly beyond the scope of this paper to uncover the reasons for this convergence in criminological interest. Nevertheless, the psychological perspective common to most of this body of work, driven by the ever-expanding literature on violence risk assessment, are chief factors that sit well within the broader priorities of modern criminal justice systems. From a dangerousness standpoint, this kind of criminological literature approaches a pathological perspective, portraying the dangerous criminal as particular in constitution, personality, upbringing and/or social background.

From a historical standpoint, however, it is often recognised that this strain of dangerousness theory can be traced to XIX century psychiatric theory and, most importantly, to the Italian criminological positivism heralded by Lombroso, Ferri, and Garofalo (Morrison 1997: 124–8). Criminological positivism insisted on the "abnormal" quality of criminals and especially of the most violent and anti-social ones. More to the point, these authors explained that crime itself is secondary, insofar as it is best understood as a symptom expressing an underlying pathology. Crime becomes in some sense contingent, less relevant than the inner nature of the offender. These points converge into a highly influential notion of criminal dangerousness, namely the contention that the danger posed by the individual, and not the crime, is the most important aspect. Crime and danger are best explained with reference to the abnormal personality of the would-be offender (Anitua 2010: 190–1).

Here we find a mostly *naturalistic* approach to danger, objectified as an intrinsic characteristic of some individuals that is more or less independent of external variables. From this point of view, several features of *periculum* come to the fore, blended in an individual who poses a risk both for himself and others. For example—and emphasising the "adventurous" dimension implicit in this etymological pathway—dangerous individuals have sometimes been portrayed as "adrenaline junkies". Their "addiction" to excitement arises from a diminished sensitivity to nervous stimulation, requiring a higher "dose" (of adrenaline) for experiencing thrills. This argument has frequently been made in reference to youth crime, but also with regard to the most paradigmatic of dangerous individuals: the psychopath (Johnson 2019: 82). From this naturalistic perspective, the offender is portrayed as inherently dangerous, leaving only preventative and defensive measures as solutions to the problem of managing the psychopathic offender.

The discussion tends to be slightly essentialist in nature and, consequently, focuses on building appropriate safeguards against such threatening individuals.

Little is said from either a relational or power-conscious perspective. Nevertheless, a kind of *dominium*-oriented interpretation could be attempted. In Foucault's view, the sovereign and the criminal (especially the high-profile criminal) bear some conceptual similarities. The two are to be seen in a genealogical relationship.[3] In sum, Foucault argues that both are despots, "the sovereign above the law and the criminal below it" (Foucault 2001: 90). Both behave like tyrants; one does so towards a victim, the other towards society at large. In a more restricted sense, the serious offender, be he sexual, violent, both, or otherwise, puts himself in a relationship of domination with the victim, where he temporarily becomes the sovereign. Foucault's, however, is hardly standard criminological theory.

3.2 Historical objections to the standard theory: the example of poverty and *dominium*

Seen in a Foucauldian light, the issue of dangerousness differs significantly from the aforementioned "standard" theory. From a historical standpoint, a kind of monolithic and individualistic vision of criminal dangerousness, such as described above, falls far short of revealing the whole picture. Human societies have feared many things and in many different ways and at different times. The fear of other human beings is certainly no exception. I will now provide some examples to illustrate this claim.

If we take the modern era[4] as a broad chronological period, one of the most stable constructions of dangerousness is the *dangerous poor*. The creation of a large class of "dangerous poor" had multiple interlocking causes, including the rural exodus that accompanied the urban concentration of wealth, and the weakening of traditional social networks. The hyper-growth of poverty, far beyond society's capacity to absorb, contributes to the ideological transformation of the poor from sanctified to demonised. The framework of the demonised poor was, above all, a projection of the fears and anxieties of the rich and powerful, who

---

[3] Actually, in Foucault's analysis of monstrosity in criminological imagination one can appreciate a blending of the naturalistic and social frameworks of *periculum* and *dominium*: "what defines the monster is the fact that, in its own existence and form, not only it is violation of the laws of society, but also of nature" (Foucault 2001: 57).
[4] Much of what is described here can be seen from the sixteenth century onwards. Nevertheless, in the nineteenth century literature the association of poverty and danger seems to peak in intensity, as it intertwines with serious political instability and economic agitation. This is especially visible in France, where many of the historical insights discussed here come from.

felt surrounded by the threatening presence of those who had nothing and so feared nothing.

As brittle as its foundations were, this proved to be a durable way of thinking about the poor, as it reinforced a status quo built on the obligation of work and the criminalisation of vagrancy, begging and idleness. Urban industrialisation changed some of these dynamics, but did not rupture the association between poverty and dangerousness. The so called "dangerous classes" came to represent the social, political, and even epidemiological, nightmare of social commentators (Nye 1984: 62).

Returning to our analytical starting point, here we see *dominium* in a much clearer light and yet, somehow, inverted. We see a construction of danger that was rooted in the social distribution of power. But here the label of dangerousness does not apply in any way to the powerful. On the contrary, the label of dangerousness is put at the service of the powerful to disqualify those whose existence is characterised, precisely, by powerlessness. Danger, naturalised in the sense of *periculum*, and ascribed to an *essentially dangerous* group of people, appears as an effect of *dominium* that simultaneously prolongs and hides *dominium*. Here we can see that the success of certain constructions of dangerousness has much to do with the discursive privileges of the groups promoting these constructions. Nevertheless, there is, no doubt, one sense in which the poor were seen as having some sort of mesmerising power over the wealthier classes. The poor were innumerable, and, deprived of everything, they had nothing to lose. This seemed to make the rich and powerful tremble, for their position was precarious and could be lost at any time if the poor were to band together against them.

Here, the power of the poor was indistinguishable from the power of the masses, which was one of the great topics of the XIX century in revolution-prone France. In this sense, political considerations have also been significant motivating factors in the discourse of dangerousness. For Gustave LeBon, writing in the late XIX century (2002, 2013), it was quite clear that the political strength of the masses was to be feared just as much as their irrational behaviour. On the other hand, for the 1789 revolutionaries a century earlier, and for Robespierre especially—before he erected his infamous guillotines (Erlenbusch 2015: 194)—there was no greater source of danger than the tyrant (Robespierre [1794] 2016; Žižek 2009). It is clear that the discourse of dangerousness is influenced by political ideology.

As far as our framework is concerned, French revolutionary discourse understood *dominium* quite well, and its political struggle on that basis. But at some point during the XIX century, the main variables for political action changed significantly and, with those changes, revolutionaries were deprived of the possibility of overpowering the State, either by words or by force. The first experiments in mass-democracy in France seemed to dissolve sovereignty into the common will and, in so doing, they hid *dominium* in plain sight. Once more, this allowed for the inverted use of *dominium*. And when, at the end of the nineteenth century, anarchist terrorism began to strike down heads of state across Europe (even if it failed more than it succeeded),[5] it was represented by the media, institutions and the public as an unfathomable evil that could bring down nations if vigorous defences weren't activated. Individual and disorganised as it was, anarchist terrorism was dressed in *dominium*, so to speak, creating the illusion of a destructive power which it probably not possess (Jensen 2004: 139). The label of dangerousness was an effect of power, one that allowed power to be exercised.

3.3 "Social" and "critical" insights on the dangerousness label

It is challenging to see just how much the examples provided above have fed into criminological thought about the theory of dangerousness. However, these examples are useful in inspiring sociological and critical approaches to the label of dangerousness.

The term "label" itself is a sign of a significant departure from the standard theory of dangerousness. From a critical, sociological point of view, the aforementioned naturalistic approach on criminal danger is theoretically limited, insofar as it fails to put the dangerous subject into a broader context of social relations. Naturalism thus reifies the dangerousness label, and makes the political and social context seem unproblematic, because it simply takes these for granted.

This critique of naturalism encapsulates both dimensions of the discussion—the sociological contention and the critical one. For the sociologically inclined, whatever the features of any given individual may be, they should at least be understood within an environment made up of social relations and institutions. Even if taken at face value, the kind of fear inspired by the terrorist or the vagrant (for example), cannot be grasped on the basis of their individual characteristics. It is not their hypothetical abnormality, but the specific position they occupy within the social structure, and the way in which they exert tension

---

5 The complexities of this first wave of terrorism are described in the works of Jensen. Suffice it to say that, from the second half of the nineteenth century onwards, the first experiments on modern terrorism can be observed. While Russia gets involved in a political struggle that would end up in revolution during the first decades of the twentieth century, in Europe and the United States we see a serious moral panic developing together with the first strikes and bombings directed at heads of state, high bureaucrats and sometimes also civilians. Expressively, the time between 1892 and 1901 has been called the "Decade of Regicide".

within that structure, that is paramount. In a sociological sense, being feared by others is at least as important as being inherently threatening.

Moreover, to see dangerousness as a label introduces another layer of complexity, insofar as it then becomes a *definition*, and the label itself an act of social communication. But communication requires more than an individual in a vacuum. Someone must be at the emitting and at the receiving end of the message, and there must be an adequate means of transmission. Moreover, for labelling to be effective it must reach significant sectors of a population, in order to create a collective representation that then affects social interactions. Ultimately, this turns attention to the different roles that actors and institutions have in spreading a message (or a label), mediated by their position and influence in the social fabric. At that point, it is important to question who is supposedly at risk and whose interests are at stake.

In historical terms, all this has tended to give the advantage to the social elites in defining what is to be feared. Although this should not necessarily be taken as a universal feature of the discourse of dangerousness, especially in more pluralistic contemporary societies, it suggests that counterhegemonic definitions of danger face considerable hurdles before they can reach comparable diffusion. This is the case especially when they are met with popular resistance. The struggle to obtain wide diffusion might best be understood, at least in some cases, as competition among the powerful for the ideological support of the masses. It is among the masses that culturally hegemonic definitions can become part of what is socially considered "common sense," where they can be hard to counterbalance. For instance, the widespread fear of terrorism, although historically tied to a preservation of institutional legitimacy and the general *status quo*, reached its peak *through* popular support rather than against it. The fear of the migrant, as an alien version of the dangerous poor, can be understood in this same way, as we shall discuss below.

The inverse of this economy of discourse lies, again, in a reinstatement of the ancient intuition underlying *dominium*. One must examine the social distribution of power and use this to contrast common sense definitions of danger. More importantly, one should not forget to pay attention to the inherent danger *of* power, which arises from the increased ability of the powerful to inflict harm more widely and more deeply. In criminology, this general view has a long, if somewhat heterogeneous, tradition, and can be seen as underlying the sub-fields of *Critical Terrorism Studies*, *Social Harm* and *Zemiology* scholarship, as well as a sector of *Green Criminology*, and *State-Corporate Crime* research, to name only a few.

Nevertheless, these are relatively unsuccessful ways of using the label of dangerousness. Despite the seemingly obvious contention that one should fear what can indeed cause harm, social constructions of danger tend to follow those frames that have been in use since at least the XIX century: the violent subject, the abnormal/mentally ill individual, the migrant/foreigner or racialised minority, or the poor. In most cases, this is certainly perplexing, since the assertion of their dangerousness often belies their actual harmfulness from a material standpoint. The historical exaggeration of the capabilities of terrorist organisations to topple governments, for example, has not weakened the linkage of terrorism with the realm of existential threats faced by the state. Accordingly, terrorism is often framed as warfare, from which perspective it is sometimes portrayed as an "asymmetric" form (Erlenbusch 2018: 149). While this could be a contentious subject, the perennial representations of extreme danger posed by the murderer, the psychopath, the "sexually violent predator" and the like, contrasts strongly with their statistical rarity and the limited harm they in fact cause, in societal terms (Seddon 2008: 308–9). With respect to the "dangerous poor," the question to ask is: how could it be that the most populous socioeconomic strata (the relatively poor), could be considered, societally, a significant source of danger? Migrants, moreover, those not even living in their country of origin, with all the existential burden and vulnerability this entails, could hardly be more dangerous than the native poor. To put it differently: how could a relative majority, the poor, be a threat, in democratic terms? And how could a vulnerable minority be a significant threat, in material terms, to the majority?

The preliminary conclusion of this criminological reflection reinforces the view that the figure of the dangerous individual is, in many ways, a representation that draws on a few, very effective, social symbols. Labelling people as dangerous is, above all, a form of social communication.

4 Dangerousness discourse in times of migration

With the theoretical ground established, it is time to turn to a specific aspect of the wider issue at hand. In the following section I will discuss the interlocking of dangerousness discourse, migration, and criminalisation. In this case, insights gained from the etymological branch tied to *dominium* will be of particular interest. As we shall see, reintroducing the question of the social distribution of power is needed to adequately tackle the "dangerous migrant" frame. This is necessary from a critically oriented perspective, one aimed at deconstructing the frame (or knocking it down altogether). Even from a strictly analytical point of view the labelling of the migrant as dangerous cannot be separated from the social distribution of the power to confer meaning. The question of why we fear the powerless can hardly be given the common-sense consecration it so frequently receives.

4.1 The dangerous-vulnerable outsider

It is widely recognised that the "dangerous outsider" framework is closely tied to human communities and their boundaries. Historically, again, this is exemplified by the evidence that strangers coming from beyond the walls inspired fear in small, remote feudal communities. Here, the tight social network that constituted the basis for integration and security (Castel 1997: 38) made everything that was unknown, and everything that came from "outside," frightening. The French term *sans aveu*[6] seems expressive of such fears, as it served to name those who were not known to the community, those who could not be recognised or answered for by a vassal or a lord and thus, could not claim the protection of the latter. In its modernisation, the term has shifted its emphasis to the idea of a person that no one would want to recognise, while retaining its overtones of nomadism and homelessness (*sans feu ni lieu*[7]). More recently, and by extension, it has been used as a derogatory term to designate someone of bad reputation, with a distasteful and immoral way of life: *sans foi ni loi*, meaning 'without faith nor law'.

In present times, it is surprising how little has changed. When it comes to descriptions of migration and human mobility, the insinuations of criminality are never far away. In fact, the *irregular*[8] nature of much of the mediatic migration (which is *a priori* an administrative status),[9] has become functionally synonymous with a criminal offense, whether actual or expected (Brouwer et al. 2017: 110; Zedner 2019: 326). This close association between migration and delinquency in the social common sense is consistent with a specific understanding of social

---

[6] *Sans aveu* can be translated as 'without confession', and should be understood in its historical sense, as discussed in the main text.
[7] Expression that could be translated, somewhat literally, as: 'without home/hearth/fire nor place'.
[8] For the purposes of this paper, irregularity should be understood as *administrative irregularity*, such as being undocumented, for instance. This is important when discussing migration, as it tends to clarify that what is often called *illegal* migration is not a criminal offense, despite being frequently treated as such. A clear example of the criminalisation of migration lies in US legislation, analysed in Chapter 5 in this volume. See also the discussion in Chapter 4 on the terms used by the US Supreme Court to describe irregular migration.
[9] Here, drawing again from the French, one would be tempted to consider the present category of the *sans papiers* ('undocumented') as a bureaucratic translation of the original *sans aveu*. Nowadays, the regular relationship with the state (the administrative status) has become the main prerequisite for inclusion, however formal that inclusion might be. In the context of human mobility, the relationship between states becomes a highly important variable and, while the human rights framework has struggled to overcome this state-centric definition of human worth and personhood, the past decades of border control and migration crises attest to the difficulty of the task.

control. From the perspective of those who would frame migrants and the poor as "dangerous", people lacking ties with the community are implicitly seen as being better placed to "take the gamble" that crime represents, because, unlike the citizens, they have "nothing to lose." Conversely, formal and informal social control are assumed to be strong enough to disincentivise criminal activities in the ordinary citizen. This neglects the fact that many crimes depend upon successful social integration.

Additionally, when class considerations are added to this picture, a kind of "intersectional" construction of dangerousness can be seen, although the interpretation of economic need is often ambivalent. On the one hand, and coherently with the *periculum* framework, a great deal of migration is framed as brave and desperate behaviour which is simultaneously astonishing (because of the *risks* it often entails) and understandable (because of the empathy elicited by desperation). In this view, one sees the migrant as a *risk-taker*, facing perils to achieve a better life, but at the potential cost of losing everything. However, the *dangerousness* attributed to the migrant comes from that same economic desperation. Desperation can seem to entail that the migrant will do "whatever it takes" to reach a vital goal, without consideration of others or of potential costs. For the truly desperate, risk can cease to be a check upon criminal behaviour. Desperation is, in a sense, a situation in which risk no longer guides behaviour.

Nevertheless, the contradiction between positive and negative constructions of the migrant, which portray the migrant as both *endangered* and *threatening*, is easily resolved when the context is explained. The migrant as risk-taker faces many hurdles on his journey, but most of those hurdles arise precisely from the opposition of the receiving community, which understands the desperation and yet fear what it entails. Such fear comes from knowing that the community itself can be one of the obstacles to overcome, a wall to be knocked down. Sometimes, this fear will manifest as identification with the State's efforts to combat irregular migration, which will be assumed to be legitimate and fair. On other occasions, this solidarity will be absent, but the citizen will still fear becoming collateral damage of the general migration policy.

Interestingly, *criminal* dangerousness is not necessarily the most important aspect of the matter here. Although the migration-crime framework is alive and well, another strong construction of danger comes from the perspective of potential economic harm (Boza Martínez 2020: 11). Here, the integrated citizenry fears the degradation of its own life conditions as an effect of the economic pressure of the most desperate in a context of limited resources. This is one of the strongest conceptions that often drives apart the "local poor" from the "migrant poor" (Jiang & Erez 2018: 9). Nevertheless, the economic frame also inspires fear in other social strata that are not in direct competition for resources with the "outsiders", such as the middle and upper classes who fear illegal occupation of their real estate properties.[10]

Unsurprisingly, most of these proxies of dangerousness find their origin in stereotypes that are not supported by evidence. Research has shown repeatedly that migrants tend to be better law-abiders than "locals" (García España 2019: 200; Jiang & Erez 2018: 6). Perhaps because their precarious situation can be reactivated at any point, irregularity tends to condition migrants toward conformity. The prevalence of squatting (illegal occupation of property) seems similarly exaggerated,[11] and fears of squatting are statistically unjustified (Anderson 2018: 18–9). Finally, the issue of "competition for resources", even when restrained to the labour market, is unwieldly and exposed to many complexities. While the evidence remains inconclusive, some researchers contend that illegal migration has little to no effect on native wages, while others even claim it has a positive effect (Card 2009; Chassamboulli & Peri 2015).[12] In any case, this does not change the fact that social intuitions are rarely "evidence-based". Regarding the topic of this chapter, the label of dangerousness appears to have far more sway than any hard evidence on the risk of effective harm.

4.2 The vulnerable-dangerous society

As suggested above, labelling the migrant as a threat should be understood as an effect of power relations. In terms of dangerousness, this pernicious labelling

---

10 The scarcity of data regarding illegal occupation of real estate is at odds with its media coverage, where one can find explicit connection made between "illegal" migration and occupation from time to time. See for example this article on the journal *El Mundo*, titled "Illegal migration and 'squatters', two challenges that must be confronted". https://www.elmundo.es/opinion/editorial/2020/12/11/5fd25b43fdddff63918b4586.html (accessed 26 April 2023).
11 This is not to deny that the issue exists, but to criticise its instrumentalisation in media coverage. However, according to some estimates, "squatting" could affect a 0.3% of the existing dwellings in Spain. From that 0.3%, 8 out of 10 concern banks and similar entities, while the rest would be conventional owners with an unspecified number of properties. The fear of illegal occupations of inhabited dwellings stands in stark contradiction with available criminal statistics. Apart from the article cited above, see also "Una dudosa alarma sobre los okupas", published in *El País*. https://elpais.com/espana/2020-09-05/una-dudosa-alarma-sobre-los-okupas.html (accessed 26 April 2023).
12 Both references concern the US context. Debates on evidence notwithstanding, it is the frame of the discussion that often poses problem, as it portrays migrants as an economic force that is desirable when profitable and threatening when competitive. While cynical, this outlook also has the effect of avoiding any critical examination of the structural determinants of the labour market. This is visible, for instance, when the expansion of precarious and semi-formal employment is taken for granted, while the blame for the economic hardship of the working classes is placed on the migrant population.

illustrates the danger posed by the established community to the outsider—not the other way around. The manifestations of this "endangerment" are many, but we will focus here on a few specific symbolic and material dangers. From a symbolic point of view, dangerousness labelling increases the possibility of stigma, desolidarisation and dehumanisation.[13] These are symbolic mechanisms that, in turn, increase exposure to material harm through increased precariousness, social marginalisation and vulnerability to violence or criminal exploitation (Brandariz & Mendiola 2019: 10; Urbina & Peña 2019: 3). From a symbolic perspective, there is a paradoxical power-play at work here, one in which actually vulnerable groups are invested with a threatening aura, which turns the social distribution of forces on its head: the less powerful groups become harbingers of social catastrophe (through anomie and crime, or through an ill-defined economic dislocation) and the more powerful are presented as vulnerable and exposed, trapped in a kind of siege mentality[14] (Jiang & Erez 2018: 6). The efficacy of this symbolic inversion is not complete, however, until it is used as a legitimising instrument to activate measures of "social defence".

At this point, it is necessary to examine the material side of the equation, not only to give an idea of the importance of the uses of language in processes of social definition, but to bring *dominium* back into view with greater clarity. In this regard, the *crimmigration* side of the story provides useful insights into how administrative and punitive powers can be the mechanisms that allow us to see the "dangerous" nature of domination.

When thinking about migration, the exercise of power can be seen in two distinct ways, the active and the omissive. Here one could pursue Giorgio Agamben's reflection on the bond that ties sovereignty[15] to those under its protection and, more importantly, to those *outside* its protection. This bond should not be thought of as present in the first case and absent in the second. Rather, it changes shape in each instance. In the extreme cases where no institutional relationship ties the sovereign to his subject, it is "abandonment" that comes to define the relationship between them (Agamben 2006: 109). And yet, abandonment is not the absence of power, but a strong effect of power itself, one in which the subject becomes

---

13 See Chapter 8 for a detailed account on the subtleties of hate speech pragmatics and, more broadly, the way hate speech works at the intersection of symbolic and material harm.
14 This phenomenon finds a very peculiar example in the conspiracy theories analysed in Chapter 7.
15 Whatever the sovereign might be in XXI century societies of the "Global North" is a question open for debate but, for the sake of argument, let's say the sovereigns of today are sovereign States.

especially vulnerable and existentially exposed. When the rules of the *polis* no longer apply, only unrestricted power remains, be it in the active sense of *making* suffer, or in the omissive sense of *letting* suffer.

Let me bring this discussion down to earth, with some concrete examples. It is broadly recognised that foreign and migrant populations (as well as racialised nationals), especially those from precarious socioeconomic situations, tend to be overrepresented within the prison system (Blanco Cordero & Doval Pais 2013: 52; Brandariz García & Fernández Bessa 2017: 310–314; Jiang & Erez 2018: 13). Although these statistics can be taken to support the view that a subject's lack of community belonging removes disincentives towards crime, I would argue that imprisonment rates are better seen as a condensing metric for multiple variables feeding into each other and composing a picture of biographic precariousness. In a certain sense, the prison system intervenes at the end of a long chain of pressures and social hurdles. As such, migrant prison population statistics serve as an indicator of a phenomenon that is much larger than crime and criminality as such, and they should not be taken as a simple descriptor of criminal behaviour. Many other factors complete the picture of the selective *endangerment* to which migrant populations are exposed.

Characteristically, the vulnerability of the migrant *as migrant*[16] comes from the struggle with the border. The border facing the migrant should be understood not just as a geographical border, but as a socially enforced border. In this light, the experience of the border includes every hardship found along the way, from the trespassing of the confines of the country of origin to the effective entry into the country of destination, and even beyond that point. The idea of the border as a socially enforced space is, as we said, not so much geographical as political. As such, the border is best understood in terms of population management and the exercise of power (Brandariz & Mendiola 2019: 10). This is evident, for instance, in the operational flexibility with which such enforcement is applied: sometimes, the border is represented by the imaginary line embodied by border control agents and police (Martínez Escamilla & Sánchez Tomás 2019: 31); on other occasions, when several consecutive fences are installed at the borders, only the innermost one is considered the *real* border, in order to argue that no crossing has taken place until every *epidermic* layer of the sovereign state has been crossed, so to speak (Martínez Escamilla et al. 2014: 6). In its increasing abstraction, the border becomes *mobile* to some extent, which means, from a legal perspective, that

---

16 That is to say, not taking into consideration the histories of precariousness that originate well before mobility is engaged. This is an extremely important variable, however, as it determines the posterior decisions on migration itself, as well as the type of hardships anticipated before, during and after crossing the border.

the applicability of the legal framework of the receiving country can also be seen as advancing or retreating, sometimes chasing the migrant, sometimes escaping him. Again, the sovereign binds or abandons, but sovereign power is always exerted.

This is not the only way to perceive border control as an expression of *dominium*. It is not necessary to examine the frequent claims of disproportionate violence exerted by authorities (already quite a significant phenomenon) to recognise the ample discretionary powers involved in policing the border. Discretionary powers are prone to the most paranoiac features of human intuition, those which feed into racism and irrational misconstrual of risk (criminal or otherwise) (Van Der Woude 2022: 46). Border control officers' self-perception as human extensions of the State's crime-prevention strategy can strengthen the construction of the migrant as a threat to the well-integrated citizen. Officers' self-perception can thus promote questionable uses of their discretionary powers (Van Der Woude & Brouwer 2017: 174). Finally, the "frontier experience" sometimes also ends in migration detention centres, places which often act as denaturalised prisons. That is to say, the experience ends in places designed to detain people, yet without the procedural safeguards and material means present in ordinary prisons (Menjívar et al. 2018: 4; Sainz 2019: 298). In such cases, the incarceration of the outsider is like a caricature of the incarceration of the "ordinary" offender, showing that the migrant's administrative *irregularity* can be a lower status than ordinary *criminality*.

Our list of examples could end with a brief exposition of some of the ways in which the border "sticks" to the migrant and, more generally, of the ways the construction of the migrant as a threatening individual creates a context of institutionalised suspicion that enables further manifestations of *dominium*. On the one hand, it is a classic contention that administrative irregularity activates in the subject an existential dread that stems from precariousness. In this sense, the border becomes internalised: it follows the migrant as a dormant condition that could be reactivated at any point (Jiang & Erez 2018: 14). This in turn exerts a strong pressure on marginalised migrants and opens the door to other forms of domination, such as economic and employment exploitation (Ferrajoli 2019: 187; Jiang & Erez 2018: 15). Several things could be noted here: precariousness and marginality increase vulnerability which, at the same time, creates further opportunities for domination and this, in turn, further spreads vulnerability. With this expansion of "opportunities" comes a decentralisation of domination, which is no longer exerted through institutional outlets, but is distributed throughout the social fabric. Consequently, any would-be exploiter is turned into a potential "small-scale sovereign," with the ability to threaten the most basic human rights and vital necessities of the precarious migrant. In addition, the institutional "abandonment" that derives from the impossibility of reuniting their protective and "immunological" functions for the migrant means that individuals cannot be effectively protected by the same authorities that could have them expelled.

Lastly, if by any chance the fears of the ordinary citizen were to be confirmed, and the migrant were pushed into crime, the dangerousness label would be reactivated in the manner of a self-fulfilling prophecy. By committing crime, even if pressed to do so through marginalisation, the outsider becomes what he was expected to be all along: ill intentioned, opportunistic and ungrateful (Jiang & Erez 2018: 12). From that point on, the criminal justice system takes the reigns, sometimes "helped" by racial profiling (Fernández Bessa 2016: 5). At this point, two options arise: on the one hand, to expel the offender back to his country of origin, often with no concern for the specific circumstances of the case (Contreras Román 2019: 328); if the option of expulsion isn't available however, we face what Bhatia calls a "bitter irony" for the punitive system: "to serve its ultimate goal of keeping out the 'undesirables', it in turn pulls them inside the managerial black hole" (2020: 41). That is, it puts them into prison.

As can be seen, then, the frontier is a social institution that, when combined with irregularity and dangerisation, creates a systematic accumulation of hurdles that are, effectively, an instrument of power. Taken together, they generate a strong pressure for conformity that produces a more easily exploitable population. And if the migrant were to become integrated with the community, he would have done so at a much higher cost than those already integrated by default. It is virtue, not to say submission, that is enforced as the *de facto* standard of adoption for the migrant. This is closely related to label of dangerousness: it is only through the irrefutable proof of innocuity and docility that the fear of the other can be dispelled. And it is only through the power of *dominium* that such docility is tested.

5 Discussion

In this chapter, we have examined the general topic of the entanglement of language with the realities it represents, particularly in legal and criminological thought, while looking closely at some of the specific ways this entanglement is manifested in contemporary debates on migration and human mobility across national borders. From a theoretical and linguistic standpoint, the notion of danger (and dangerousness) has been used as the axis for the analysis. To conclude, some key points can be outlined.

Etymologies do not necessarily have any influence on the current meaning of words. Nevertheless, etymologies are useful as conceptual time capsules that contain traces of historical meanings which, in turn, can be helpful in conceptual reasoning and in the interpretation of current meanings and usages of words. Starting from this premise, we examined two different etymological branches related to the notion of danger—*periculum* and *dominium*—which bear several distinct features. The etymological trace-meanings of both of these words are still implicit in contemporary uses of the word *danger*, but they are rarely given equal importance. Broadly speaking, there is a more naturalistic or abstract vision of *danger*, used for describing things or possibilities, that stands in contrast with a characteristically social vision that portrays *danger* as an effect of power relations.

As the criminological excursion in this paper has made explicit, both broad notions have accompanied historical criminology, and are still implicit in contemporary criminological scholarship. It could even be argued that behind these competing definitions of danger lies a general divide that has tended to separate more "orthodox", naturalistic views on crime and deviation from critically oriented ones. From the first perspective, inherently dangerous individuals have to be identified, studied and controlled. From the other, the selection of dangerous individuals is always flawed because it is blind to the social effects of the uneven distribution of power and, as such, is barely able to effectively prevent social harm.

Here, we chose to connect this discussion with the idea of the label, conceiving the discourse of dangerousness as an exercise in social communication, one aimed at making some definitions of danger prevail over others. From this point of view, *dangerisation* (portraying someone as dangerous) and *endangerment* (putting someone in actual danger) must be understood as separate but interconnected. This paper has focused on the migration debate, an area in which such a distinction leads to some paradoxical results.

With this in mind, it is possible to see how, in spite of the efforts of legal scholarship and international organisations, the human rights framework often fails to account for the situation of the migrant. The recent migration and/or refugee crises illustrate quite clearly that legality is strongly at odds with political priorities. This has put state institutions and authorities in a strange and Janus-faced position: obliged to enforce legality, while often tinkering with the conditions for its applicability. In so doing, governments have often shown that bureaucratism can be a useful tool to hide xenophobic border management policies and practices under a democratic and civilised façade. Recent history shows, however, that some administrations have not even tried to engage in such *double jeu*, which demonstrates how far racism, securitarianism and infantile conceptions of cultural identity have intertwined into a new fascistic knot.

The classic portrayal of the migrant poor as a multidimensional threat is a kind of dangerousness discourse that has proven quite resilient, both to statistical counterevidence and ethical reasoning. From the framework adopted here, it is possible to see that the discourse of dangerousness is not only a word game, but a power-play made through words. Here, vulnerable populations are portrayed as a threat by members of secure and established populations, who see themselves, in turn, as the vulnerable targets. This allows for the unleashing of preventive and defensive measures, while deactivating human solidarity or even the most basic humanitarian consensus. Dangerisation opens the door to endangerment, through action or omission, which shows that labelling populations as dangerous can be (and maybe is by definition) simultaneously an effect of power and a way of exerting power. Successful labelling can indeed be extremely useful, as it facilitates population control and, in some cases, submission and exploitation. This can happen through the coordinated assembly of administrative powers and policing, but also through the diffusion of exploitation throughout the social fabric, enabled by marginality, clandestinity and vulnerability.

All this can be said to describe a well-established social practice, and yet it must be underlined that it is supported by a mainly linguistic operation: labels and the definitions of words both feed public perceptions and articulate overt or implicit practices. However, language is also the main instrument of legal discourse, which shows the importance of revisiting, reviewing and revising the juridical underpinnings—our language itself—when necessary, because these underpinnings should serve as a stable common ground when tackling difficult matters. This is not a mere manifestation of a desire regarding present societies: the migration crisis has occasionally shown that, despite an unfavourable political and social climate, legality has helped contain some excesses and stem the xenophobic tide, at least to some extent.[17] But behind the efficacy of legal discourse lies an active struggle, because the curtailment of human rights cannot (should not) gain ground on the basis of communicative pirouettes. The struggle for a culture of legality and human rights is, in a sense, a struggle over the legitimate uses of language. As such, contemporary debates cannot be dismissed without the risk of institutions being further permeated by dangerisation, securitisation and xenophobia.

---

17 See Chapter 9 for a discussion of how Spanish criminal policy has been designed to *protect* vulnerable populations and, significantly, migrants.

References

Agamben, Giorgio. 2006. *Homo Sacer*. Valencia: Pre-Textos.
Anderson, Miriam. 2018. Landlords and squatters: Neither victims nor executioners?: Context and outline of the Spanish reform on eviction of illegal occupiers. *Red-Revista Electronica De Direito* 17 (3). 14–38.
Anitua, Gabriel Ignacio. 2010. *Historias de los pensamientos criminológicos*. Buenos Aires: Editores del Puerto.
Baudrillard, Jean. 1981. *Simulacra and simulation*. Detroit: University of Michigan Press.
Berger, Peter L. & Thomas Luckmann. 1967. *The social construction of reality: A Treatise in the sociology of knowledge*. London: Penguin.
Bhatia, Monish. 2020. Crimmigration, imprisonment and racist violence: Narratives of people seeking asylum in Great Britain. *Journal of Sociology* 56(1). 36–52.
Blanco Cordero, Isidoro & Antonio Doval Pais. 2013. La extranjería como factor de discriminación de los condenados a penas de prisión. In Luis R. Ruiz Rodríguez, *Crimen organizado y extranjería en España y Marruecos*, 51–84. Valencia: Tirant lo Blanch.
Boza Martínez, Diego. 2020. Crimmigration as electoral strategy: The case of Brexit. *Revista Española de Investigación Criminológica* 18(2). 1–30.
Brandariz García, José Ángel & Cristina Fernández Bessa. 2017. "Perfiles" de deportabilidad: El sesgo del sistema de control migratorio desde la perspectiva de la nacionalidad. *Estudios Penales y Criminológicos* 37. 307–347.
Brandariz, José Ángel & Ignacio Mendiola. 2019. Introducción: Una mirada interdisciplinaria a la frontera como espacio y proceso. *Revista CIDOB d'Afers Internacionals* 122. 7–14.
Brouwer, Jelmer, Maartje van der Woude & Joanne van der Leun. 2017. Framing migration and the process of crimmigration: A systematic analysis of the media representation of unauthorized immigrants in the Netherlands. *European Journal of Criminology* 14(1). 100–119.
van Brunt, Brian. 2015. *Harm to others: The assessment and treatment of dangerousness*. Alexandria, VA: American Counseling Association.
Campbell, Jacquelyn C. & Jill Theresa Messing. 2017. *Assessing dangerousness: Domestic violence offenders and child abusers*. New York: Springer.
Card, David. 2009. Immigration and inequality. *American Economic Review* 99(2). 1–21.
Castel, Robert. 1997. *La metamorfosis de la cuestión social*. Buenos Aires: Paidós.
Chassamboulli, Andri & Giovanni Peri. 2015. The labor market effects of reducing the number of illegal immigrants. *Review of Economic Dynamics* 18(4). 792–821.
Cohen, Stanley H. 2015. *Demonios populares y "pánicos morales": Delincuencia juvenil, subculturas, vandalismo, drogas y violencia*. Barcelona: Gedisa.
Contreras Román, María. 2019. La expulsión administrativa como excusa de inacción penal. *Revista Crítica Penal y Poder* 18. 319–332.
Corominas, Joan. 1987. *Breve diccionario etimológico de la lengua castellana*. Madrid: Gredos.
de Vaan, Michiel. 2008. *Etymological dictionary of latin and the other italic languages*. Leiden: Brill.
Erlenbusch, Verena. 2015. Terrorism and revolutionary violence: The emergence of terrorism in the French Revolution. *Critical Studies on Terrorism* 8(2). 193–210.
Erlenbusch, Verena. 2018. *Genealogies of terrorism: Revolution, state violence, Empire*. New York: Columbia University Press.
Fernández Bessa, Cristina. 2016. Transformaciones de la penalidad migratoria en el contexto de la crisis económica: El giro gerencial del dispositivo de deportación. *InDret* 4. 1–25.

Ferrajoli, Luigi. 2019. Políticas contra los migrantes y crisis de la civilidad jurídica. *Revista Crítica Penal y Poder* 18. 182–193.
Foucault, Michel. 2001. *Los anormales*. Madrid: Akal.
Foucault, Michel. 2004. *Nietzsche, la genealogía, la historia*. Valencia: Pre-Textos.
García España, Elisa. 2019. Más inmigración, menos delincuencia. *Revista Crítica Penal y Poder* 18. 194–205.
Harrison, Karen. 2011. *Dangerousness, risk and the governance of serious sexual and violent offenders*. London: Routledge.
Hoad, T. F. 2000. *Oxford concise dictionary of English etymology*. Oxford: Oxford University Press.
Jensen, Richard Bach. 2004. Daggers, rifles and dynamite: Anarchist terrorism in nineteenth century Europe. *Terrorism and Political Violence* 16(1). 116–153.
Jiang, Jize & Edna Erez. 2018. Immigrants as symbolic assailants: Crimmigration and its discontents. *International Criminal Justice Review* 28(1). 5–24.
Johnson, Scott A. 2019. Understanding the violent personality: Antisocial personality disorder, psychopathy, & sociopathy explored. *Forensic Research & Criminology International Journal* 7(2). 76–88.
Koopman, Colin. 2013. *Genealogy as critique: Foucault and the problems of modernity*. Bloomington: Indiana University Press.
LeBon, Gustave. [1895] 2002. *The crowd: A study of the popular mind*. Mineola, NY: Dover Publications.
LeBon, Gustave. [1894] 2013. *The psychology of revolution*. Aukland: The Floating Press.
Martínez. Escamilla, Margarita & José Miguel Sánchez Tomás. 2019. La vulneración de derechos en la frontera sur: De las devoluciones en caliente al rechazo en frontera. *Revista Crítica Penal y Poder* 18. 28–39.
Martínez Escamilla, Margarita, José Miguel Sánchez Tomás, José Luis Segovia Bernabé, José Luis Díez Ripollés, Elisa García España, Enrique Gimbernat Ordeig, Julio V. González García, Esteban Pérez Alonso & Mercedes Pérez Manzano. 2014.*"Expulsiones en caliente"*: Cuando el Estado actúa al margen de la Ley. Informe Jurídico, 27 de junio de 2014.
McSherry, Bernadette & Patrick Keyzer. 2011. *Dangerous people: Policy, prediction and practice*. Abingdon: Routledge.
Menjívar, Cecilia, Andrea Gómez Cervantes & Daniel Alvord. 2018. The expansion of "crimmigration," mass detention, and deportation. *Sociology Compass* 12(4). 1–15.
Morrison, Wayne. 1997. *Theoretical Criminology: From modernity to post-modernism*. London: Cavendish.
Nye, Robert A. 1984. *Crime, madness and politics in modern France: The medical Concept of national decline*. Princeton: Princeton University Press.
Partridge, Eric. 1966. *Origins: A short etymological dictionary of modern English*. London: Routledge.
Rennie, Ysabel Fisk. 1978. *The search for criminal man: A conceptual history of the dangerous offender*. Lexington, MA: Lexington Books.
Robespierre, Maximilien. [1794] 2016. *Virtud y terror*. Madrid: Akal.
Sainz, Pablo. 2019. La memoria de los CIE. *Revista Crítica Penal y Poder* 18. 297–308.
Sarbin, Theodore R. 1967. The dangerous individual: An outcome of social identity transformations. *The British Journal of Criminology* 7(3). 285–295.
Seddon, Toby. 2008. Dangerous liaisons: Personality disorder and the politics of risk. *Punishment and Society* 10(3). 301–317.
Stumpf, Juliet P. 2006. The crimmigration crisis: Immigrants, crime, and sovereign Power. *American University Law Review* 56(2). 367–420.

Urbina, Martin Guevara & Ilse Aglaé Peña. 2019. Crimmigration and militarization: Policing borders in the era of social control profitability. *Sociology Compass* 13(2). 1–16.

Woude, Maartje van der. 2022. Getting to the core of crimmigration: The imagined reality that is Schengen. *Kriminologisches Jorunal* 43(1). 36–52.

Woude, Maartje van der & Jelmer Brouwer. 2017. Searching for illegal junk in the trunk: Underlying intentions of (cr)immigration controls in Schengen's internal border areas. *New Criminal Law Review* 20(1). 157–179.

Zedner, Lucia. 2019. The hostile border: Crimmigration, counter-terrorism, or crossing the line on rights? *New Criminal Law Review* 22(3). 318–346.

Žižek, Slavoj. 2009. *Sobre la violencia*. Barcelona: Paidós.

Monika S. Schmid
Chapter 3
Language attrition as a problem for language analysis for the determination of origin

1 Introduction

When asylum seekers arrive in the country where they plan to seek refuge, they often do not have the required documentation, such as birth certificates or passports, to prove the truth of their claimed national origin and their persecution. In cases where the narrative is in doubt, many countries use the method of Language Analysis for the Determination of Origin (LADO, cf. Patrick 2012). The method consists of an interview conducted by a native or proficient speaker of one (or more) of the languages the refugee can be expected to speak based on their claimed origin, along with an analysis of this interview by someone with native and/or specialised knowledge of that language. The fundamental assumption underlying the LADO practice is thus that "if someone has spent a substantial part of his [sic] life in a certain area, in particular his youth (where early language acquisition takes place) he may be expected to speak at least one of the language varieties that are commonly used in that area".[1] The view expressed here is one that is widely held: people grow up learning and speaking the language variety or varieties of their environment, and once they have become native speakers of that variety or varieties, they will not only remain identifiable as such for the rest of their lives but will also be able to readily determine if someone else is or is not a native speaker.

---

1 "Taalanalyse gaat uit van de volgende vooronderstelling: als iemand het grootste deel van zijn leven in een bepaald gebied heft doorgebracht, en met name zijn jeugd (waarin vroege taalverwerving plaatsvindt), mag verwacht worden dat hij ten minste één van de taalvarianten spreekt die in dit gebied gangbaar zijn". (Vakbijlage taalanalyse, Ministerie van Justitie, Immigratie- en Naturalisatiedienst ct. after Zwaan ed. 2008.) English translation in main text is mine.

Note: This chapter is a slightly adapted version of a previously published chapter, Schmid, M. S. (2019). Language attrition as a problem for LADO. In Patrick, Schmid & Zwaan, K. (eds.) *Language Analysis for the Determination of Origin* (pp. 155–165). New York: Springer, reproduced here with kind permission of the publisher.

Monika S. Schmid, University of York

It has often been pointed out that the assumption that LADO is something native speakers can do with ease is problematic (for an overview, see Patrick et al. eds. 2019). Many factors can complicate this task, for example when speakers grow up in diverse multilingual contexts (Muysken 2019) or when listeners operate under preconceived notions (Preston 2019) Questions have also been raised concerning the expertise and qualifications necessary to reliably perform such a task (Fraser 2019).

A further complicating factor relates to the stability of the native speaker status. While the possibility that the native language (L1) can change when the speaker resides in a different linguistic environment—a process known as *first language attrition* (Schmid 2011, 2013)—is highly relevant in the complex context of refugees, it has so far received little attention in the literature and research surrounding LADO. The notion that knowledge, use and proficiency of a native language can change under certain circumstances—that is, that native speakers may effectively cease to be native speakers after lengthy exposure to other languages or varieties—is one that is often perceived to be strange or even ludicrous. For instance, when the American soldier Bowe Bergdahl was released after five years of Taliban captivity in May 2014, and his father subsequently announced to the press that he had "trouble speaking English", this caused a flood of incredulous and derisive comments on Twitter and in the media.[2] The view expressed most frequently was that native speakers do not forget their first language (L1), and certainly not after a period of only a few years.

A substantial body of research conducted over the past few decades has demonstrated, however, that the knowledge, proficiency and production of an L1 are not entirely stable and unchangeable entities. Findings from migrants who have lived immersed in a second language (L2) environment show that all levels of language are susceptible to change and influence from other languages that a speaker learns, knows and uses.

Language attrition is a process of linguistic change and development common among migrants who live in a second language environment. L1 attrition has been shown to affect all linguistic levels, from a reduction of productive vocabulary (Yılmaz & Schmid 2012; Schmid & Jarvis 2014) through an increase of disfluency and hesitation phenomena in spoken language (Schmid & Beers Fägersten 2010) to morphosyntactic phenomena such as case marking (Schmitt 2010) or pronoun use (Chamorro et al. 2016). The most relevant finding in the context of LADO, however, is that attriting populations are consistently perceived to be less native-like than monolinguals residing in the country of origin (e.g. de Leeuw et al. 2010; Bergmann et al. 2016).

---

2 https://languageattrition.org/celebrity-attriters/ (accessed 12 August 2022).

Of particular interest here is a study by Hopp and Schmid (2013), which compared advanced L2 learners of German having either English (n = 20) or Dutch (n = 20) as their L1 (all speakers were late bilinguals who had acquired German after the age of 10 and had lived in Germany for a minimum period of three years) with native German speakers who had emigrated either to Anglophone Canada (n = 20) or to the Netherlands (n = 20) and had lived there for an extended period of time (9 years or longer). As a baseline, 20 native speakers of German living in Germany were also included.

The study used short extracts (of 12–20 seconds)[3] from a narrative task in which participants had either described a complex picture or narrated a film sequence. The extracts were played to 130 native speakers of German (first-year students at the University of Mannheim), who were asked to determine whether the person that they had just heard was a native speaker of German or not. These ratings were elicited in two steps: the first question was a yes/no question ("Is this person a native speaker of German?"), and the second was a three-point confidence rating. This led to an effective 6-point Likert scale, with a 1 indicating certainty that

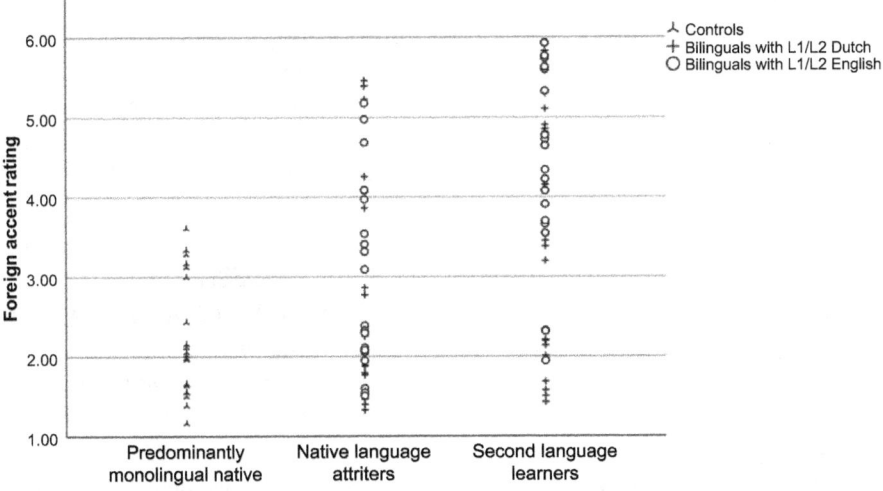

Figure 3.1: Foreign accent ratings of 40 L1 attriters and 40 L2 learners of German (L2ERS) in comparison with 20 predominantly monolingual native speakers of German (Hopp & Schmid 2013, Figure 5, used with permission of CUP).

---

3 While the duration of these samples may not seem very long, most linguistic investigations of foreign accent use clips of similar length, and it has been established that it is enough to make a determination (Schmid & Hopp 2014). In fact, one paper suggests that native speakers can make accurate determinations after as little as 30 ms (Flege 1984).

the speaker was a native German, and a 6 indicating confidence that the sample was not native speech.

The results from this experiment are displayed in Figure 3.1. This graph illustrates that the raters were not entirely unanimous even where the Germans in Germany were concerned: in some cases, people appear to have been somewhat uncertain as to the speaker's native speaker status. None of them, however, fell into the non-native range, which starts at 4 on this scale. On the other hand, a fair number of the L1 German migrants were rated to be probable or certain foreigners, while several of the second language learners fell clearly within the unambiguously native range.

The above graph suggests that the L1 attriters cluster more towards the lower (the native) end of the scale, and that more of the L2 learners are at the upper (the foreign) end. However, it is also clear that these results are not categorical: both of the bilingual groups are spread out over more or less the same range of perceived accents. These findings strongly indicate that even people who left the country where they grew up as adults cannot reliably be distinguished from foreigners once they have spent a certain amount of time abroad.

The situation becomes yet more unstable where people who emigrated in childhood or adolescence are concerned. It has often been found that children who emigrate in the company of their parents, or who are born in the country of migration but brought up speaking a different language at home, typically fail to attain native standards in the home language, performing much more like second language learners. Often, they have comparatively low—sometimes only receptive—proficiency, even though parents have made every attempt to enforce the use of the "native" language at home (see e.g. Montrul 2008; Schmitt 2010).

These assumptions were confirmed in research conducted by Karayayla and Schmid (2019), who compared four groups of L1 speakers of Turkish (n = 30 in each group): predominantly monolingual Turkish speakers who had lived in Turkey their entire life, late migrants who had emigrated at an average age of 22 years (range: 11–42), early migrants who had emigrated at age 9.5 (range 7–11) and speakers who were either born in the UK or came there before age 3. Karayayla and Schmid elicited foreign accent ratings for all of these speakers according to the same procedure described for Hopp and Schmid (2013). The findings from this experiment are presented in Figure 3.2.

Again, these findings show that even among the adult migrants there are some speakers who are no longer perceived as natives. However, among the younger migrants and the second-generation speakers, foreign accents are found much more consistently. There was, moreover, a strong correlation between the age of migration and the perceived foreign accent (r = .56, p < .001). These findings suggest that

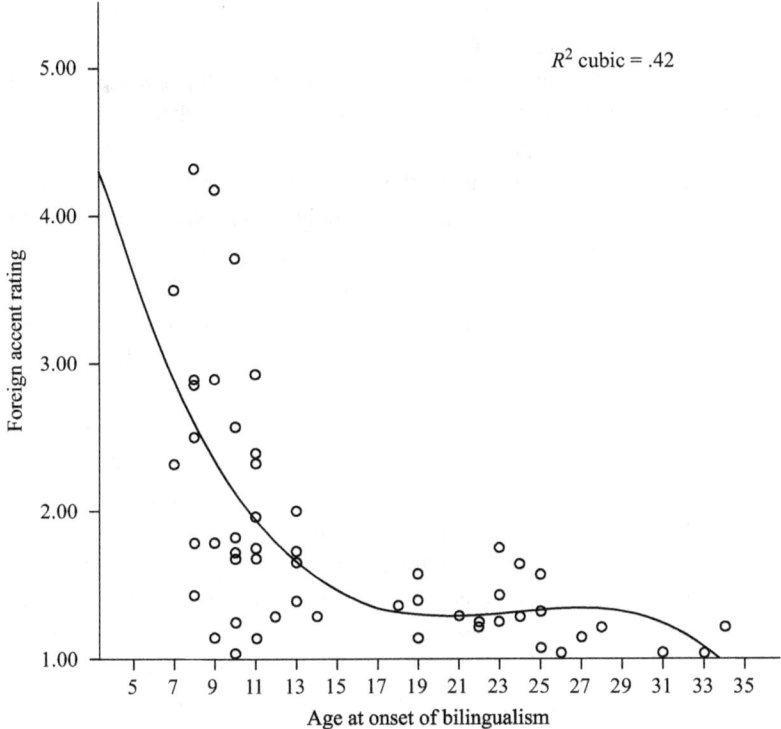

Figure 3.2: Scatterplot of the relationship between age at onset of bilingualism and foreign accent rating (Karayayla & Schmid 2019, Fig. 4, used with permission of Wiley Publishing).

younger speakers who acquire a second language may never attain anything like a stable native speaker status in their L1.

The most striking findings in this respect come from an investigation of international adoptees conducted in Paris by Christophe Pallier and his team (Pallier et al. 2003). Pallier et al. studied a group of young adults who had been born in Korea but had been adopted by French parents when they were between the ages of 3 and 9. In this study, the researchers were unable to find *any trace* of the mother tongue. For example, they made them listen to the sequence of numbers from one to ten in various different languages that were unknown to them, such as Japanese, Polish, Wolof, among others, also including Korean. The task for the participants consisted merely in indicating whether the language they had just heard was Korean or not. The adoptees failed at the task: they did not perform any better than monolingual French speakers who had never been exposed to Korean.

Pallier and his colleagues also conducted fMRI scans of the adoptees' brains while they listened to different languages, and here, too, they did not find any specific activity elicited by Korean as opposed to other, entirely unfamiliar languages. The mother tongue of these speakers had apparently been entirely erased from their brains. Subsequent studies of international adoptees have substantiated these findings: when adoptees attempt to re-learn their birth language, they have few advantages over speakers who have never been exposed to this language. The only area in which such adoptees were found to have a slight advantage concerns the perception and production of some phonemes that are particularly challenging for L2 learners (Hyltenstam et al. 2009; Oh et al. 2010), but they do not extend to lexicon and/or morphosyntax. The implication that early exposure may affect purely phonetic perception has also been validated by a recent fMRI study of Chinese adoptees in Francophone Canada, which finds that these speakers recruit the same brain areas when hearing tones in nonsense syllables that play a role for Chinese-French bilinguals, and both populations are different from French monolinguals with no knowledge of Chinese (Pierce et al. 2014).

The deterioration of the first language can take place astonishingly fast, as longitudinal investigations of international adoptees, such as the one presented by Isurin (2000), suggest. Isurin reports a case study of a Russian child adopted (at age 9) by American parents who asked the researcher (a native Russian speaker) to provide regular interaction with the child in her L1. Notwithstanding these attempts, Isurin reports a rapid breakdown of first language proficiency, which after a relatively short period (around one year) was followed by a total refusal of the subject to interact in Russian with the investigator. In general, extremely rapid loss (over a space of months or even weeks) of the ability and willingness to use the L1 appears to be a hallmark of international adoption (e.g. Glennen & Masters 2002; Nicoladis & Grabois 2002; see also the overview in Hyltenstam et al. 2009).

Similarly, German Jews who were rescued from Nazi Germany on the so-called *Kindertransporte* after the pogrom of 9 November 1938 and placed in English-speaking foster families recall losing German so quickly (within months or even weeks) that correspondence with their parents who had remained in Germany became severely impaired—as is recounted, for example, in the documentary film *Into the Arms of Strangers* and in autobiographies such as Edith Milton's *The Tiger in the Attic* (2005).

2 Implications for LADO

The findings from investigations into L1 attrition have real and serious implications for LADO, in particular since this tool is commonly used in many countries to assess the origin of children from the age of 4 years and up (ten Thije 2008), and is also used to re-evaluate the claims of asylum seekers who have in the past been granted asylum but about whose case histories some doubt has been raised (de Munnik 2008).

Consider the case of the Dutch politician and writer Ayaan Hirsi Ali. Hirsi Ali was born in Somalia in 1969. From an early age, her life was characterised by flight and dislocation. Since her father was considered a dissenter, he was arrested and thrown into prison when Hirsi Ali was still a child. Her mother, siblings and Hirsi Ali herself fled the country when she was age 6. Together they spent years in exile in Saudi Arabia, Sudan and Ethiopia, before eventually settling in Kenya in the 1970s, where Hirsi Ali received most of her schooling. In 1992 she was to be sent from Kenya to Canada to enter into an arranged marriage. To escape her forced marriage, Hirsi Ali made her way via Germany to the Netherlands, where she applied for asylum. Her status as an asylum seeker initially condemned her to living in a caravan park and to doing unskilled labour. Upon her arrival she spoke no Dutch, but she quickly became fluent (and eventually near-native) in that language. Receiving Dutch citizenship in 1995, Hirsi Ali studied politics and within ten years had become one of the most prominent and outspoken members of parliament. Her meteoric rise to success and her controversial books have made her internationally known. In 2006, Hirsi Ali was recruited by the American Enterprise Institute (a neoconservative think-tank) and she relocated to the United States (Buruma 2006; Hirsi Ali 2007).

In 2006 it became public knowledge that Hirsi Ali had given false information on her asylum application, changing both her name and her date of birth in order to make it more difficult for her family members to find her and force her into the arranged marriage. A rather undignified media storm ensued around the question of whether she should lose her citizenship. Although Hirsi Ali was eventually allowed to retain her Dutch citizenship (and, to my knowledge, no LADO analysis was ever conducted on her), hers was exactly the kind of position many migrants find themselves in: years after their arrival, an inconsistency may be detected in their original applications, whether due to actually inaccurate information having been provided at the time or due to misunderstandings in the original interviews. In these situations, LADO is one of the standard mechanisms used to resolve problems posed by the conflicting information. A negative LADO verdict is taken as evidence of an untruthful account, often resulting in the loss of the refugee's residency permit.

Refugee biographies are rarely straightforward journeys from A to B. In many cases the speaker left the country of origin—and thus the linguistic environment where the L1 was spoken—a considerable time before. S/he may no longer be perceivable as a native speaker. LADO language analysis may therefore not provide reliable results, in particular if long periods of this time were spent in countries where similar languages or varieties are spoken.

In this context it would be valuable to know during which period language attrition takes place. To date, however, this has not been established, although there are indications that the bulk of linguistic change may happen fairly early on, within the first few years after migration (Schmid 2011). What has been shown is that, after a migration span of 10 years or more, length of residence no longer correlates with attrition phenomena—which is good evidence that attrition is a fairly rapid process (Köpke & Schmid 2004; Schmid 2011), even among adults. For children, as was seen above, the process of loss can typically be measured in months rather than years.

3 Traumatisation

One last issue must be mentioned, pertinent to the case of perceived nativeness and linguistic deterioration of the first language among asylum seekers; namely, the impact of traumatisation. There is considerable psychological evidence that emotional factors play an important role in memory formation and recall, particularly as they relate to processes such as repression/suppression and distortion, as opposed to processes that occur as a result of non-use, such as memory decay and slowdown or failure of retrieval, or interference processes due to competition between two languages (Ecke 2004). Such explanations are often—and convincingly—invoked in the cases of international adoptees discussed above. However, where attrition among adult migrants is concerned, the picture to date is less clear. Available findings, rather counterintuitively, suggest that language attrition is usually only very slightly affected by external factors, such as the frequency of use of the language in question or the attitude towards the native and host language and culture (e.g. Schmid 2013). A series of large-scale empirical investigations of the attrition of a range of native languages, investigating linguistic habits and language and cultural attitudes in detail, has largely failed to find correlations between how often a speaker uses his or her first language, how they feel towards it, and how they perform on a wide range of linguistic tasks (e.g. Keijzer 2007; Schmid 2007; Dostert 2009; Cherciov 2010; Yilmaz & Schmid 2015). However, it should be noted here that all of these studies investigate migrants who left their country of origin voluntarily, usually either for reasons of career or family/relationship.

There is one notable exception to this pattern, namely the investigation of first language attrition, use and maintenance among German-Jewish refugees who escaped Nazi persecution in the 1930s, reported by Schmid (2002). In this study it was shown that the extent of traumatisation that an individual suffers can impact heavily on the maintenance or deterioration of the first language. Those speakers investigated who had experienced events such as the pogrom of 9 November 1938 performed significantly worse on all measures of nativeness than those who had left relatively soon after the Nazi seizure of power, and thus had not been traumatised to the same extent. The results of this study thus suggest, in keeping with psychological theories of memory and repression/suppression/distortion, that a genuine asylum seeker—who, by definition, will have experienced trauma and hardship—may come to be perceived as less native-like than someone with less traumatic memories. It is evident that such findings are highly relevant in the context of language analysis as a determinant of origin.

4 Practical examples of the problems of language analysis

The fact that (potential) language attrition is not a factor that is generally recognised as problematic for LADO implies that such analyses may routinely be conducted in cases for which they are not suitable. This has led to a breakdown of the procedure or, in the worst cases, to unfair deportations. It is impossible to determine exactly how many, and with what outcomes in each case. That it has occurred, on the other hand, is demonstrable on the basis of a number of cases in which I have been asked, by the lawyer of the asylum seeker in question, to provide an expert opinion.

To illustrate, I will shortly summarise some of those cases here.

a) A 30-year-old Somali-born male left Somalia at age 10 in the company of his mother. Both lived in Saudi Arabia for some years, and quickly switched to speaking Arabic even among themselves. As an adult he did not speak nor understand Somali. The speaker came to the Netherlands aged 27, and his failure to use Somali was initially interpreted as evidence that his account was false. The official documentation asserted that, given his history, Mr. A should be a proficient speaker of Somali. Upon reviewing his case, my conclusion was that the loss of the native language in this specific case should by no means be considered exceptional or surprising. It was of vital importance for both son and mother to acquire Arabic as quickly as possible, because both had to work among other speakers of Arabic and had no remaining ties to Somalia (the speaker's grandparents and father having all passed away). Under such circumstances, a child will lose the native language almost by necessity.

This assessment was accepted by the court where Mr. A had appealed against the original rejection of his application, and he was eventually granted asylum.

b) In a series of other cases, I was asked to assess the reliability of analyses conducted on speakers who had left their country of origin years before, upwards of age 8 and often aged more than 20. Some of these speakers were adolescents or young adults at the time of departure, others were children. Again, the official documentation indicates that the prevalent view is that children who escape together with their families should acquire their parents' dialect, not whatever variety of the language may be spoken in the different places where they reside. That this is not the case is amply demonstrated by the findings of Karayayla and Schmid (2019) discussed above—children do not grow up to become perfect linguistic copies of their parents, instead they pick up whatever language or variety is spoken by their peers.

c) The last case is of a slightly different nature. It concerns a 17-year old boy who came to the Netherlands from Angola at age 9 and was taken in by a Dutch foster family. He made a huge effort to learn Dutch quickly and well, but completely forgot his native Portuguese. Upon his reaching adulthood, he was faced with the prospect of extradition, against which he and his foster family appealed on the grounds that it is unethical to deport a 17-year old, alone, to a country where he does not speak the language. The official argument, made among others by the Minister of Integration (and echoed in the press, particularly among letters from the readers), was that this claim was probably a lie, as, given his history, he should have a native command of Portuguese.

5 Conclusion

The problems of the change and deterioration that may be observed in the native language of a speaker living in a different linguistic environment are of immediate relevance and concern to the practice of LADO. It is imperative that immigration authorities be made aware of these findings, and that safeguards should be implemented. When it comes to speakers who leave the country where they grew up after puberty, they can no longer be reliably identified as natives after a period abroad, which may be as short as 3 to 5 years. No LADO analysis should be attempted after this timespan has elapsed. Younger children, even if they emigrate in the company of their parents, often experience a drastic loss of their first language, and children who emigrate alone almost invariably forget it completely within months. These findings should inform the practice of language analysis to determine the origin of asylum seekers, and should be used to assess whether or not such an analysis can indeed provide a reliable result in any individual case.

References

Bergmann, Christopher, Amber Nota, Simone A. Sprenger & Monika, S. Schmid. 2016. L2 influence causes non-native-like pronunciation of vowels and consonants in German L1 attriters. *Journal of Phonetics* 58(1). 71–86.

Buruma, Ian. 2006. *Murder in Amsterdam: The death of Theo van Gogh and the limits of tolerance*. London: Penguin.

Chamorro, Gloria, Antonella Sorace & Patrick Sturt. 2016. What is the source of L1attrition?:The effect of recent L1 re-exposure on Spanish speakers under L1 attrition. *Bilingualism: Language and Cognition* 19(3). 520–532.

Cherciov, Mirela. 2010. *Between attrition and acquisition: The case of Romanian in immigrant contexts*. Toronto: University of Toronto dissertation.

de Leeuw, Esther, Monika S. Schmid & Ineke Mennen. 2010. The effects of contact on native language pronunciation in an L2 migrant setting. *Bilingualism: Language and Cognition* 13(1). 33–40.

de Munnik, T. 2008. Het gebruik van de taalanalyse in het asielrecht: afschaffen, bijschaven of continueren? [the use of language analysis in asylum law: discard, adapt or continue?] In Karin Zwaan (ed.,) *De taalanalyse in de Nederlandse asielprocedure: Een juridische en linguïstische verkenning* [language analysis in dutch asylum procedure: a legal and linguistic exploration], 25–72. Nijmegen: Wolf Legal Publishers.

Dostert, Susan C. 2009. *Multilingualism, L1 attrition and the concept of 'native speaker'*. Düsseldorf, NRW: Heinrich-Heine-Universität Düsseldorf dissertation.

Ecke, Peter. 2004. Language attrition and theories of forgetting: A cross-disciplinary review. *International Journal of Bilingualism* 8(3). 321–354.

Flege, James Emil. 1984. The detection of French accent by American listeners. *The Journal of the Acoustical Society of America* 76(3). 692–707.

Fraser, Helen. 2019. The role of native speakers in LADO: Are we missing a more important question? In Peter L. Patrick, Monika S. Schmid & Karin Zwaan (eds.), *Language Analysis for the Determination of Origin*, 71–89. New York: Springer.

Glennen, Sharon & M. Gay Masters. 2002. Typical and atypical language development in infants and toddlers adopted from Eastern Europe. *American Journal of Speech-Language Pathology* 11(4). 417–433.

Hirsi Ali, Ayaan. 2007. *Infidel: The story of my enlightenment*. New York: Free Press.

Hopp, Holger & Monika S. Schmid. 2013. Perceived foreign accent in L1 attrition and L2 acquisition: The impact of age of acquisition and bilingualism. *Applied Psycholinguistics* 34(2). 361–394.

Hyltenstam, Kenneth, Emanuel Bylund, Niclas Abrahamsson & Hyeon-Sook Park. 2009. Dominant-language replacement: The case of international adoptees. *Bilingualism: Language and Cognition* 12(2). 121–140.

Isurin, Ludmila. 2000. Deserted island or a child's first language forgetting. *Bilingualism: Language and Cognition* 3(2). 151–166.

Karayayla, Tugba & Monika S. Schmid. 2019. L1 attrition as a function of age at onset of bilingualism: L1 attainment of Turkish-English bilinguals in the UK. *Language Learning* 69(1). 106–142.

Keijzer, Merel. 2007. *Last in first out?: An investigation of the regression hypothesis in Dutch emigrants in Anglophone Canada*. Amsterdam: Vrije Universiteit Amsterdam dissertation.

Köpke, Barbara & Monika S. Schmid. 2004. First language attrition: The next phase. In Monika S. Schmid, Barbara Köpke, Merel C.J. Keijzer & Lina Weilemar (eds.), *First language attrition: Interdisciplinary perspectives on methodological issues*, 1–2. Amsterdam: John Benjamins.

Milton, Edith. 2005. *The tiger in the attic: Memories of the Kindertransport and growing up English*. Chicago: University of Chicago Press.

Montrul, Silvina A. 2008. *Incomplete acquisition in bilingualism: Re-examining the age factor*. Amsterdam: John Benjamins.

Muysken, Pieter. 2019. Language and origin: The perspective of multilingualism. In Peter L. Patrick, Monika S. Schmid & Karin Zwaan (eds.), *Language Analysis for the Determination of Origin*, 119–130. New York: Springer.

Nicoladis, Elena & Howard Grabois. 2002. Learning English and losing Chinese: A case study of a child adopted from China. *International Journal of Bilingualism* 6(4). 441–454.

Oh, Janet S, Terry Kit-Fong Au & Sun-Ah Jun. 2010. Early childhood language memory in the speech perception of international adoptees. *Journal of child language* 37(5). 1123–1132.

Pallier, Christophe, Stanislas Dehaene, Jean-Baptiste Poline, Denis LeBihan, A.-M. Argenti, Emmanuel Dupoux & Jacques Mehler. 2003. Brain imaging of language plasticity in adopted adults: Can a second language replace the first? *Cerebral Cortex* 13(2). 155–161.

Patrick, Peter. 2012. Language analysis for determination of origin: Objective evidence For refugee status determination. In Lawrence M. Solan & Peter M. Tiersma (eds.), *The Oxford Handbook of Language and Law*, 534–546. Oxford: Oxford University Press.

Patrick, Peter, Monika S. Schmid & Karin Zwaan (eds.). 2019. *Language Analysis for the Determination of Origin. Current Perspectives and New Directions*. New York: Springer.

Pierce, Lara J., Denise Klein, Jen-Kai Chen, Audrey Delcenserie & Fred Genesee. 2014. Mapping the unconscious maintenance of a lost first language. *Proceedings of the National Academy of Sciences* 111(48). 17314–17319.

Preston, Dennis R. 2019. Trouble in LADO-Land: How the brain deceives the ear. In Peter L. Patrick, Monika S. Schmid & Karin Zwaan (eds.), *Language Analysis for the Determination of Origin*, 131–154. New York: Springer.

Schmid, Monika S. 2002. *First language attrition, use and maintenance: The case of German Jews in Anglophone countries*. Amsterdam: John Benjamins.

Schmid, Monika S. 2007. The role of L1 use for L1 attrition. In Monika S. Schmid, Barbara Köpke, Merel C.J. Keijzer & Lina Weilemar (eds.), *First language attrition: Interdisciplinary perspectives on methodological issues*, 135–153. Amsterdam: John Benjamins.

Schmid, Monika S. 2011. Contact x time: External factors and variability in L1 attrition. In Monika S. Schmid & Wander Lowie (eds.), *Modelling Bilingualism: From structure to chaos*, 155–176. Amsterdam: John Benjamins.

Schmid, Monika S. 2013. First language attrition. *WIREs Cognitive Science* 4(2). 117–123.

Schmid, Monika S. & Kristy Beers Fägersten. 2010. Disfluency markers in L1 attrition. *Language Learning* 60(4). 753–791.

Schmid, Monika S. & Holger Hopp. 2014. Comparing foreign accent in L1 attrition and L2 acquisition: Range and rater effects. *Language Testing* 31(3). 367–388.

Schmid, Monika S. & Scott Jarvis. 2014. Lexical access and lexical diversity in first language attrition. *Bilingualism: Language and Cognition* 17(4). 729–748.

Schmitt, Elena. 2010. When boundaries are crossed: Evaluating language attrition data from two perspectives. *Bilingualism: Language and Cognition* 13(1). 63–72.

ten Thije, Jan. 2008. Taalanalyse in een meertalige asielprocedure [language analysis in a multilingual asylum procedure]. In Karin Zwaan (ed.), *De taalanalyse in de Nederlandse asielprocedure: Een juridische en linguïstische verkenning*, [language analysis in Dutch asylum procedure: a legal and linguistic exploration] 121–145. Nijmegen: Wolf Legal Publishers.

Yılmaz, Gülsen & Monika S. Schmid. 2012. L1 accessibility among Turkish-Dutch bilinguals. *The Mental Lexicon* 7(3). 249–274.

Yılmaz, Gülsen & Monika S. Schmid. 2015. Second language development in a migrant context: Turkish community in the Netherlands. *International Journal of the Sociology of Language* 236. 101–132.

Zwaan, Karin (ed.). 2008. *De taalanalyse in de Nederlandse asielprocedure: Een juridische en linguïstische verkenning.* [language analysis in Dutch asylum procedure: a legal and linguistic exploration] Nijmegen: Wolf Legal Publishers.

Laura M. Hartwell
Chapter 4
"Once an alien has passed through our gates": Noncitizens in three US Supreme Court oral arguments

1 Introduction

This chapter focusses on the range of appellations of noncitizens by actors of the United States Supreme Court, as manifested in oral arguments related to detained noncitizens. This study is the first to address the use or non-use of expressions such as *noncitizen* and *illegal alien* by prominent US jurists in a context of national and international debate surrounding such expressions (Hartwell 2021a). The objective of this study is to examine the oral references by justices and advocates during US Supreme Court arguments when referring to persons who are not US citizens. This discourse is reviewed under the lenses of both ideology and temporality. Renowned legal scholar Erwin Chemerinsky affirms that there is "a sense that it is the 'law', not the justices, that is responsible for the Court's decisions. This is nonsense and always has been. The Court is made up of men, and now finally women, who inevitably base their decisions on their own values, views, and prejudices" (2014: 10).

Ideology influences Supreme Court actors, a fact widely acknowledged in the way justices are commonly branded as either "conservative" or "liberal". As Law Professor Jeffrey Rosen (2016) writes: "Perhaps the greatest sign of Scalia's influence is that liberal justices and scholars now make arguments about constitutional text and history, insisting that the conservative justices are ignoring the text and original understanding of the Constitution". Similarly, prizewinning legal journalist Linda Greenhouse highlights that the "death of the court's leading liberal [Ruth Bader Ginsburg] and her nearly instantaneous replacement by Amy Coney Barrett, a conservative forty years her junior, jolted the flow off course, upending expectations for the term that was just getting under way" (2021: 239).

It is therefore hypothesised in the present chapter that Supreme Court discourse, stemming from divergent ideological viewpoints, will reflect the historic controversies surrounding US non-citizens. To test the hypothesis, the oral arguments of three US Supreme Court cases specifically pertaining to detained noncitizens were

Laura M. Hartwell, Université Toulouse Capitole

analysed: *Zadvydas v. Davis* (2001), *Jennings v. Rodriguez* (2017) and *Johnson v. Guzman Chavez* (2020). This chapter first offers a review of the historical practices of such English appellations in the United States, before attending to recent debates in Canada and Europe. Then, after describing the methodology of the study, three guiding questions are explored. (1) What are the general trends in the appellations of the noncitizen population by both justices and advocates? (2) How does ideology influence these references? (3) How have these appellations evolved over the last two decades? The results suggest that the absence of these jurists' use of the expression *illegal alien*, plus the influence of ideology on jurists' terminology and an evolution of the lexical references to the heterogenous population of people living without citizenship status in the US, have led to a recent acceptance of the term *noncitizen*, irrespective of jurists' ideological alignment.

2 US terminology referring to noncitizens: a diachronic view

The notion of US citizenship has historically been limited by racial and economic criteria, starting as early as the US Naturalization Act of 1790 that limited citizenship to "free white persons". The Alien and Sedition Acts of 1798 proclaimed that "all natives, citizens, denizens, or subjects of the hostile nation or government, [. . .] not actually naturalized, shall be liable to be apprehended, restrained, secured and removed, as alien enemies". This "Enemy Aliens Act" was extended to allow for the detainment of 110,000 people, including 70,000 US citizens of Japanese descent during World War II (Cole 2003). Furthermore, as Lavissière and Hill discuss in Chapter 5 in this volume, the "Undesirable Aliens Act" of 1929, which criminalized unlawful entry into the United States, was motivated by "racial animus".

The Library of Congress initiated the subject heading *Alien* in 1910, followed seventy years later by its subheading *Alien, Illegal* (1980). It was not until 1993 that *Illegal aliens* became a Library of Congress subject heading, reflecting, according to its declaration, a widespread societal use of the term:

> As with all LC subject headings, *Illegal aliens* was based on literary warrant: it was needed to describe the topics being covered in published literature, and it employed the terminology used in the literature and in reference sources. The literature addressed the legal status of persons who entered and lived in a country illegally, the social conditions under which those persons lived, and the services provided to them, including education, health, and employment, among other topics (Library of Congress, Policy and Standards Division 2016: 2).

In response to recent community movements, actors of the press have announced their refusal to relay the term *illegal* when referring to a person. Several newspapers discontinued the use of the appellation *illegal alien*, notably the *Miami Herald* (2003), *Huffington Post* (2008), *The Los Angeles Times* (2012). The Associated Press followed suit in 2013.

This public debate has also reached university settings and active student groups. In 2014, Dartmouth College modernized the subject heading in their library catalogue, adopting *undocumented immigrants* to replace the term *illegal aliens*. Other universities followed, including the University of Colorado (2018), the California State University Libraries (2020), Harvard University (2021), Middlebury College (2021) and Vassar College (2021) (Kelsey et al. 2021).

Dartmouth College librarian John DeSantis submitted proposals in June 2014 to remove *illegal alien* through the Library of Congress' Subject Authority Cooperative Program (SACO). The proposal was initially accepted by the US Library of Congress in 2016 (Baron & DeSantis 2017), before the decision was reversed. The issue remains unresolved. DeSantis' 2016 proposal to replace *aliens* with *noncitizens* and *illegal aliens* with *unauthorized immigration* as subject headings was refused by a vote in Congress, marking the only time in US history that the Congress has voted on a Library of Congress subject heading. This vote occurred in the face of clear and public disapproval by library professionals: "The delay in the decision from the Library of Congress to revise this heading is unfortunate and the library community should not lose sight of the real pain and alienation that having this offensive terminology in our catalogues can entail for members of a marginalized community" (Fox et al. 2020).

Harvard Library also recently distanced itself from the Library of Congress, refusing the subject heading *illegal alien* and adopting the subject headings *noncitizen* and *undocumented immigrant*. Rebecca Martin, Scholarly Communications Librarian of Harvard University's Gutman Library explains: "Intentionally not using the Library of Congress subject heading is a really defiant act in a way, [but] our task force feels really strongly that our users, especially our students, should not see pejorative terms describing them from the University" (*The Harvard Gazette* 2021).

Legislators have responded in the wake of this public attention. The New York City Human Rights Law (NYCHRL) established guidelines under the Enforcement Guidance on Discrimination on the Basis of Immigration Status and National Origin (2019), according to which, "The use of the terms 'illegal alien' and 'illegals', with the intent to demean, humiliate, or offend a person or persons in the workplace, amounts to unlawful discrimination under the NYCHRL" (2019: 4). (For an in-depth analysis of the dangers of "hate speech", see Chapter 8 in this volume). At a national level, the US Immigration reform bill HR1177 introduced in 2021 would facilitate the obtention of citizenship or resident status for certain *undocumented individuals* by creating a new status of *lawful prospective immigrant*. It would also specifically replace *alien* with *noncitizen* in federal US immigration statutes (Sánchez 2021). Thus, the question of terminology is woven into Federal immigration policy.

3 International responses to the use of the term *alien*

The US press and the Library of Congress, whose subject headings are frequently adopted by non-US libraries, have exported the term *alien*. However, automatic foreign language translators, such as Google Translate or DeepL, often construe the word *alien*, particularly when found in sentence fragments, to refer to an extra-terrestrial being. By contrast, these same translators will often propose the expression *foreign national* rather than *alien* when translating from another language into English. Thus, at the international level, the distinct US use of the term *alien* is not well apprehended by automatic translators nor, in consequence, by their international users (Hartwell 2021b).

Governmental institutions have also considered and rejected the term *alien*. In 2013, the European Parliament withdrew the term *alien* from documents relative to *Eurodac*, a European Union-wide biometric database that collects and maintains fingerprint records of non-EU/European Economic Area (EEA) nationals and asylum applicants. During the Parliamentary debates, British member of the European Parliament Robert J. E. Evans stated: "In English 'aliens' are people from another planet and to be perfectly honest we do not have much of a problem with that at the moment" (Evans 1999). The President of the European Commission of Justice and Home Affairs, António Vitorino concluded that the Commission can accept amendment No. 1 (as proposed by Alima Boumediene-Thiery in commission) "to avoid the term 'alien' and instead refer to 'third country nationals', although we will have to make it clear that the Regulation also covers stateless persons" (Vitorino 1999).

A debate also arose in the Canadian Standing Committee on Citizenship and Immigration, where, in 2008, the Associate Professor of Sociology, Luin Golding (2008) testified that the term *undocumented* does not always define the many foreign nationals who initially enter Canada with authorized documentation. However, it is "a response to such terms as 'illegal alien', which have the effect of dehumanizing and criminalizing people" (Golding 2008).

4 Methods and tools

The three US Supreme Court cases most relevant to detained noncitizens were first identified: *Zadvydas v. Davis* of 2001 (Oyez n.d.a), *Jennings v. Rodriguez* of 2017 (Oyez n.d.b) and *Johnson v. Guzman Chavez* of 2020 (Oyez n.d.c). Spanning nineteen years, the transcripts of three hours of oral arguments comprise the voices of sixteen different justices and seven advocates. All of the transcripts for the oral arguments of these cases were found on the free-use Oyez (Oyez, n.d.d) multimedia archive maintained by the University of Cornell's Legal Information Institute, Justia and Chicago-Kent College of Law since 1955. The corpus analysis toolkit AntConc, created and made freely available by Anthony Laurence at Waseda University (Laurence 2019) was used for the concordancing and analysis of the transcripts.

Each case represents one hour of speech, divided in half for each party, who is regularly questioned and commented on by the justices. *Zadvydas v. Davis* (2001) resulted in a 5-4 decision in favour of Mr. Zadvydas, a stateless person who had lived for four decades in the United States before being arrested on re-entry after attending his mother's funeral abroad. The Court decided that the statute limits the "post-removal-period detention to a period reasonably necessary to bring about that alien's removal from the United States" and "does not permit indefinite detention" (Oyez n.d.a). *Jennings v. Rodriguez* (2017) resulted in a 5-3 decision in favour of the government (Decision 2). Mr. Rodriquez had entered the United States as a child, later married a US citizen and was father to a US citizen. The Court determined that "statutory provisions allowing the detention of certain [noncitizens] during immigration proceedings do not require periodic bond hearings, nor do they impose time limits on detention" (Oyez n.d.b). Finally, *Johnson v. Guzman Chavez* (2020), with decisions in favour of the government, concerned Ms. Guzman Chavez, who, upon re-entry to the United States, was determined by an immigration judge to have a reasonable fear of torture if she were to return to her home country. The Court held that "Section §1231, not §1226, governs the detention of [noncitizens] subject to reinstated orders of removal. It authorizes detention when a person is ordered 'removed' and enters the 'removal period,' which begins on the 'date the order of removal becomes administratively final'" (Oyez n.d.c).

The transcripts were read to identify and list all terms referring to noncitizens. The transcripts and terms were treated with AntCONC to verify quantities and to tally speakers of each term. Excluded from the results were any expressions referring to a document or title, for example the title within in Section 1231, "Detention and Removal of Aliens Who Have Been Ordered Removed" as quoted by Chief Justice John G. Roberts, Jr (2020). Also excluded from the results is any occurrence of a term when referring to a citizen, for example the first use of *person* in her explanation that when "someone who overstayed a visa and encountered a nasty INS person, that person could be locked up forever without any access to a judge" (Justice Ginsburg 2001). Grammatical terms, such as *she* or *him* were not included in the results. Included in the results are all words used when a speaker "self-corrects", such as *people* and *aliens* in the utterance of Mr. Stewart (2017), "the Court has said time after time that insofar as people arriving—aliens arriving at our shores are concerned, whatever process Congress chooses to give is due process". Also included are nouns that draw upon other parts of speech, such as *arrivings* in "two-thirds for the Arrivings, won their cases even when they were detained", as spoken by Mr. Arulanantham (2017). A total of 572 occurrences of terms referring to noncitizens were found.

The ideological voting pattern for each justice was identified according to the Oyez (n.d.e) website, which also provides justices' voting pattern for each case as related to ideology and seniority. Among the sixteen justices considered in this study, several judged several cases. Justices O'Connor (Oyez n.d.f) and Kennedy (n.d.g) are often believed to provide "swing votes". Justice O'Connor's decision was of "liberal" ideology in 2001, while Justice Roberts pronounced a conservative position for both the 2017 and 2020 cases. Thus, the justices whose opinions were conservative include Justices Rehnquist, Scalia, Kennedy and Thomas in 2001; Justices Roberts, Kennedy, Thomas, Alito and Gorsuch in 2017 and Justices Roberts, Kavanaugh, Barrett, Gorsuch, Alito and Thomas in 2020. For the "liberal" decisions, we find Justices Stevens, O'Connor, Souter, Ginsburg and Breyer in 2001, Justices Ginsburg, Breyer, Sotomayor and Kagan in 2017 and Justices Sotomayor, Breyer and Kagan in 2020.

The ideological stances of the advocates were based on the status of the petitioners: advocates defending the conservative governmental position are considered as holding a conservative position, while the advocates defending the noncitizens are considered liberal. The conservative pleadings were conducted by Edwin S. Kneedler (*Zadvydas v. Davis* 2001), Malcolm L. Stewart (*Jennings v. Rodriguez* 2017) and Vivek Suri (*Johnson v. Guzman Chavez* 2020). The liberal positions were defended by Robert F. Barnard and Jay W. Stansell (*Zadvydas v. Davis* 2001), Ahilan T. Arulanantham (*Jennings v. Rodriguez* 2017) and Paul W. Hughes (*Johnson v. Guzman Chavez* 2020). In this chapter, all references to quotes by advocates and by justices refer to the oral arguments of the case of the stated year.

5 Findings and discussion

The 572 occurrences were classified into three broad groups: a) *alien;* b) *people* or *person* and c) less frequent terms. Of these three groups, *alien* and its plural *aliens* represent slightly over one-third (n = 196, 34.3%) of all the occurrences. The second major group consists of the neutral terms *people, person* and the plural *persons* (n = 162, 28.3%). All occurrences of *people* or *person* referring to a citizen were excluded from this study. Finally, terms less frequently used, ranging from the indefinite pronoun *someone* to specific professions, such as *spy*, represent 37.4% (n = 214) of the occurrences.

More than half of the 572 occurrences contained either the base *alien* or *person/people*, which like other nouns, formed multi-word units. Thirty-seven of the 196 occurrences of *alien* formed the basis of seven different two-word units: *arriving alien* (n = 11), *criminal alien* (n = 11), *dangerous alien* (n = 3), *detained alien* (n = 3), *admitted alien* (n = 2), *arrested alien* (n = 2), and *resident alien* (n = 2). More complex appellations in the form of multi-word units with a neutral base are present, such as "anyone in the immigrant status who is convicted of a felony" (Justice Souter 2001) or "individuals in removal proceedings" (Mr. Hughes 2020). However, there is not a single occurrence of the expression *illegal alien*.

The back and forth of appellations, and the need to clarify the context in which they were spoken, is apparent in Example (1) from *Jennings v. Rodrigues* (2017). Here, the conservative advocate Mr. Stewart employs *newly arriving alien* to differentiate from those who may have become integrated into their community after a longer stay. Justice Kagan, Justice Breyer and Mr. Stewart draw upon expressions of *people/person* for greater precision. Justice Kagan uses "people who are coming to the border" to clarify *new admits*, while Mr. Stewart uses *that person* to refer back to "no newly arriving alien will be allowed to enter the country until he or she has persuaded the decision-maker". Justice Breyer explains that *visitors* could be people overstaying their visitor's visa. Thus, what might be considered specific terms, such as *admits* and *visitor* often require specificity as to their meaning. Even *businessperson* is clarified in this context as being someone who has overstayed their visa.

Example 1): *Jennings v. Rodriguez* **(2017)**
Elena Kagan: Mr. Stewart, is—is your argument about the **new admits**, the **people who are coming to the border,** premised on the idea that **they** simply have no constitutional rights at all? [. . .]
Elena Kagan: "Arbitrary" means that **nobody** gave them an individualized hearing, and so **we** don't know whether **they**'re being held for any good reason. **Nobody**'s made that decision. So, usually, in **our**—you know, usually, in **our** constitutional law, **we** think that that's a problem."

Mr. Stewart: Now, I—I think, Congress, consistent with the Constitution, could have abolished parole altogether and could have said, as a categorical matter, no **newly arriving alien** will be allowed to enter the country until **he or she** has persuaded the decision-maker that the right answer ultimately is to let **that person** in. I think that would be a constitutional scheme under this Court's decisions, but Congress has historically offered parole as a form of process by—to—
Stephen G. Breyer: **Visitors** too? I mean, you know, **people** overstay their visitors' visas. And we find a **businessperson** who, in fact, has overstayed his visa.

Although *nobody* is not found in these three oral arguments when referring to a noncitizen, there are a total of 63 occurrences of similar indefinite pronouns referring to the population in question: *anybody* ($n = 1$), *anyone* ($n = 3$), *everybody* ($n = 4$), *everyone* ($n = 3$), *somebody* ($n = 13$) and *someone* ($n = 39$). Indefinite pronouns such as *someone* carry a general undetermined quality that contrasts with the legal character of *person* as found in the US Constitution. Justice Ginsburg highlights this contrast during the oral arguments of *Zadvydas v. Davis* (2001): "It's quite different when you're talking about someone who was here, who was part of the community, and who has, as you say—in your brief I think you say, yes, they are persons, and the Constitution says, nor shall any person be denied due process, so it's quite different".

While the full use of pronouns is beyond the scope of this study, they constitute both a bridge and a barrier between citizens and noncitizens. Again, in Example 1, Justice Kagan's comment, "we don't know whether they're being held for any good reason" creates a we-or-them dichotomy. This dichotomy is repeated in her conclusion, "in our constitutional law, we think that's a problem" and also by Mr. Stewart's suggestion that "until he or she has persuaded the decision-maker that the right answer ultimately is to let that person in". However, pronouns such as *she, he* and *they* can refer to anyone irrespective of their citizenship or residence. Thus, linguistically, people of all populations share common human ties through the use of common pronouns.

The interweaving of this we-them dichotomy and shared pronouns is also illustrated in Example (2) from *Zadvydas v. Davis* (2001). Here, Justice Scalia (2001) employs the pronoun *you* to address a hypothetical immigrant person spoken to by a collective *we* as in "we will deport you", thereby identifying the action of deporting. The unprecise subject *we* allows Justice Scalia to avoid naming the agent of this action of deportation. Within the same enunciation, he also employs *you* to address the justices and advocates present in the Court, "it really isn't punishment, it's—you know, that was the deal". The liberal advocate Mr. Stansell responds also by employing *you*, in this case referring to Justice Scalia, "Your Honor, it—the hypothetical you lay out". Mr. Stansell transforms Justice Scalia's *immigrant* into the more general *somebody* who could be incarcerated by a far-reaching agency, "somebody through the administrative action of an administrative agency to be put in—to be incarcerated for life".

Example 2): *Zadvydas V. Davis* **(2001)**
Antonin Scalia: Then why is it unconstitutional to say to an **immigrant**, if **you** commit a felony, **we**'re not going to put **you** in prison for life, but **we** are simply not going to let **you** back into the general populace, and **we** will deport **you** if **you** can find a place to be deported to, but otherwise **you** will be held under house arrest, not punitive, but **you** will not be allowed into the general population? Why is that lesser punishment, if you consider it that, although it really isn't punishment, it's—**you** know, that was the deal. Why is that lesser sanction unconstitutional, whereas sending the **felon** to jail for life and punitive treatment for life would not be unconstitutional?
Jay W. Stansell: Well, Your Honor, it—the **hypothetical you** lay out would allow **somebody** through the administrative action of an administrative agency to be put in—to be incarcerated for life, and this Court has never condoned such an extreme civil detention process.

Thus, the pronoun *you* allows the speaker to address varied populations, be they under immigrant status or members of the Court. The pronoun *we* also allows a speaker to avoid announcing the specific agent of a given action such as in, "we will deport you" as does the passive voice, "you will not be allowed into the general population". The you-or-we dichotomy—expressed by *you* when referring to a person present or not, a vague *somebody*, or an unspecified active *we* —is transgressed by the multiple ways of employing these pronouns and of qualifying the actual agents or concerned population.

Justice Scalia's hypothetical of Example (2) brings us back to the combined notion of immigration and illegality. In Example (2), he refers to the noncitizen as an *immigrant*. The appellation *immigrant* is found only eight times. While *immigrant* may describe a person's life experience, it does not indicate citizenship because a person may immigrate and then obtain a new nationality, unlike *alien* where the person has not obtained a new nationality. A *felony* is an illegal act. No occurrences of *felonious aliens* were found, but there are 11 occurrences of *criminal alien*. In all of these, the person has been found guilty of a crime, as in the statement of Mr. Kneedler (2001): "We think that [detention pursuant to expressed statutory authority] is a very reasonable response on behalf of the two political branches about how to deal with the severe problem of dangerous criminal aliens in our midst". However, the attested criminality of *criminal aliens* found in these oral arguments contrasts with the ambiguity of the appellation *illegal alien*, where the legality of the administrative situation of the person is in question. Thus, although *criminal alien* reverberates linguistically with *illegal alien* it has a legal substance that is absent in the appellation *illegal alien.*

The appropriate use of *illegal* is confirmed by the thirty other uses of the word *criminal.* These thirty uses often serve to differentiate between *criminal* and *civil*, such as *criminal detention* ($n = 3$), *criminal history* ($n = 3$) or *criminal trial* ($n = 3$). More so, during a discussion about whether holding a client in a detention centre would be more acceptable than a prison facility, Mr. Barnard (2001) answers: "Well, being a criminal lawyer, I've never seen [a detention centre] that looked particularly appealing to the average person, so I just can't imagine that that would be the case". We also find, referring to people, *criminal population* (n = 1) and simply *a criminal* (n = 1). In none of these examples can *criminal* be replaced by *illegal* and carry anything like the same meaning. Thus, their discourse confirms the inappropriateness of the term *illegal alien*, as *illegal* refers to a certain situation or context and not to a person.

5.1 Findings and discussion: ideology and frequencies of *alien* and *noncitizen*

To examine the link between ideology and discourse, this study reviews the terminology of each actor according to two contrary ideologies: a "conservative position" versus a "liberal position". The characteristics of these ideological discourse differences range from the frequency of certain traditional terms such as *alien* or inclusive terms such as *people* or *someone*. It is also interesting to note, as we will see below, the absence of references to humans, especially by actors representing a conservative decision, when discussing the law.

As noted, the most frequent term referring to people without citizenship status is *alien*, with 196 occurrences. Of these, 167 (85.2%) are uttered by an actor adopting a conservative position, with 29 (14.8%) considered to represent a liberal position. Over two-thirds (69.4%) of the total occurrences come from advocates of the government: Mr. Kneedler (n = 49) in 2001, Mr. Stewart (n = 50) in 2017 and Mr. Suri (n = 37) in 2020. This is in stark contrast with the advocates of the detained noncitizens: Mr. Barnard (n = 3) and Mr. Stansell (n = 2) in 2001, Mr. Arulanantham (n = 0) in 2017 and Mr. Hughes (n = 3) in 2020. Thus, the term *alien* is used over sixteen times more by those attempting to convince the Court that the respondents should remain in detention than by the advocates who find that judicial review of the detention is needed.

At first glance, there appears to be a greater equilibrium among the justices. For the 52 occurrences of *alien* uttered by the justices, 31 (59.6%) are by a justice whose final decision maintains a conservative position and 21 (40.4%) are by justices who adopt a liberal position. However, there is an evolution of proportional use by the more recently appointed justices of the Court. For example, in 2001, the conservative Justice Scalia utters *alien* six times, but he also uses forms of *person/people* (n = 16), *individual* (n = 3), an indefinite pronoun (n = 1), and three other terms. During the same case in 2001 Justice Rehnquist, with less linguistic variety, employs only *alien* (n = 4), *client* (n = 1) and an indefinite pronoun (n = 1).

In the 2017 case, no conservative-leaning justice utters *alien*, reflecting the fact that their questions centre on issues of the law instead of the people affected by those laws.

In 2020, *alien* is uttered ten times by conservative Justice Alito, six times by Justice Coney Barrett and three times by Justice Kavanaugh. Although no other appellations were found for Justice Alito, the more recently appointed justices also use the more modern term *noncitizen*. Justice Coney Barret employs *noncitizen* six times as a substitute for *alien* and uses no other term. She may qualify these, for example when she asks: "Do the regulations that govern the post-order custody review automatically kick in so that the detained noncitizen gets some sort of hearing?" Justice Kavanaugh also says *noncitizen* once along with one indefinite pronoun. Justice Gorsuch's use of such appellations is limited to one occurrence of *individual*. Thus, although there is an overall ideological balance of justices concerning the use of this term, conservative justices overall exhibit little linguistic variety, except for the notable introduction of the more recent socially accepted term *noncitizen*.

More linguistic variety among the justices is found in liberal decisions. In the 2003 case, Justice Ginsburg shows much greater linguistic variety, preferring expressions with *person/people* (n = 22), indefinite pronouns (n = 5), and a single *alien*. In the 2017 case, Justice Sotomayor employs *alien* three times, such as when she is questioning the difference between criminal detention and immigration detention: "That's basically saying that we're not a country of law, that we're a country of arbitrariness in detaining people, locking them up. [. . .] [When] you detain aliens, you put them in orange suits, they are shackled during visitation and court visits, they are subject to surveillance and strip searches, they are referred to by number, not by name". Here, Sotomayor repeatedly refers, through the details she evokes and through her use of definite pronouns, to the dehumanizing nature of immigration detention—a procedure which degrades dignity to the point of erasing the person's name. Justice Kagan says *alien* only once in 2017, then liberally uses the term *alien* ten times in 2020, notably during a rapid back-and-forth of brief questions and answers: "I guess the reason I ask is because that seems very similar to me, [. . .] [You] have a final order of removal and a pending withholding decision, and yet you're going to treat that alien under 1226 but the aliens [sic] here under 1231" (2020).

It is interesting to note that the term *illegal* is entirely absent from these three oral arguments. However, ten occurrences of three forms of *lawful* are found, including six occurrences of *lawfully* and two each of *lawful* and *unlawfully*. Of these, seven are found in *Zadvydas v. Davis* (2001), four by Mr. Kneedler, who represents the conservative governmental position, for example when he states: "[An order for removal] also terminates the person's status as a lawful, permanent resident". The other three are uttered during the same case by Justices Ginsburg ($n = 2$) and O'Connor ($n = 1$) who will adopt a liberal position. The other three occurrences are by advocates of the noncitizens, including Mr. Arulanantham who in 2017 questions whether the detention is "lawful or not", and Mr. Hughes (2020) who introduces *unlawfully* ($n = 2$). Thus, these findings privilege the term *unlawful* over *illegal* in these circumstances, and this holds true across ideological lines.

In contrast with *alien*, the term *noncitizen* is a recent adoption and is employed irrespective of ideology. *Noncitizen* was not found in the earlier 2001 case. Curiously, the vast majority ($n = 22$) of the 25 occurrences of the positive *citizen* do appear in the 2001 case. For example, Justice Souter (2001) seems to search for the word *noncitizen* without finding it: "Illegally, yes, but in the United States, and the Fourteenth Amendment does not distinguish between citizens *and others* in this respect". The other three occurrences of *citizen* are uttered by the advocate defending Mr. Rodriguez in 2017, when he explains, "we have people in our case who [. . .] have married U.S. citizens, and they have a petition. And they're going to win their case. They're just waiting for the DHS to decide the petition. And they get detained like eight months, 10 months waiting for this petition to get decided. That person has no reason to flee". In the same case, Mr. Arulanantham also introduces the complementary negative *noncitizen*, "it goes far beyond anything this Court has ever said with respect to the power to detain non-citizens. [. . .] You have the power to detain, but only if the detention is necessary to ensure that the person appears or to prevent, you know, a danger to the community". (For a discussion of the notion of "danger" see Chapter 2 in this volume.)

The remaining thirteen occurrences of *noncitizen* are found in the 2020 *Johnson v. Guzman Chavez* case, where the positive forms *citizen* and *citizenship* are absent. As mentioned above, the term is used seven times by the more recently appointed conservative Justices Barret and Kavanaugh. Kavanaugh utters: "I think you acknowledged that some significant number of noncitizens in this circumstance could be detained for several years". It is also uttered once by Justice Sotomayor and five times by the advocate of Ms. Guzman Chavez. Mr. Hughes states: "the reinstatement order makes certain critical determinations, for example, that the noncitizen is actually the same person as at issue in that prior order. Additionally, that the noncitizen unlawfully entered the United States". Following his lead, an *illegal alien* could be more appropriately referred to as a "noncitizen who unlawfully stayed in the country". Overall, the use of *noncitizen* has recently crossed the ideological lines demarcating these oral arguments, unlike the term *alien*.

5.2 Findings and discussion: ideology and lexical range

The contrast of terminology in correlation with ideology is broader than the mere use of the term *alien*. The total 571 occurrences of reference to noncitizens were equally balanced between actors identified as having a conservative position ($n = 274$) and those with a liberal position ($n = 287$). However, the term *alien* was more than six times more frequent among actors holding a conservative position ($n = 167$) than for the liberal group ($n = 29$). Conservative justices uttered *alien* 31 times or 18.6% of the total conservative position occurrences, while justices adopting a liberal position uttered *alien* 21 times or 72.4% of the total liberal position occurrences. Thus, this difference in use lies essentially with the conservative advocates. In all, almost two-thirds (60.9%) of the appellations used by the conservative position actors are *alien*, which leaves little room for greater lexical variety that might encompass the many heterogeneous situations pertaining to detained noncitizens.

Liberal position actors, by contrast, employed more than twice as many expressions with *person* or *people* ($n = 119$), compared to their counterparts ($n = 44$). Although there was only a slight difference for indefinite pronouns (respectively, $n = 35$; $n = 22$), those holding a liberal position decision employed *individual* three times more often ($n = 38$) than their counterparts ($n = 12$). Thus, while the overall totals for actors manifesting a conservative or a liberal position are equivalent, the liberal group more frequently employs expressions that humanise the population in question.

Also apparent is the reduction over time of lexical diversity among both ideological parties. This is first noticed in the quantity of terms per case, which declines from 255 items in 2001, to 179 in 2017, to only 138 items in 2020. In the 2001 *Zadvydas v. Davis* case, excluding forms of *alien* and *people/person* and indefinite pronouns, fourteen different appellations are found. The justices and advocates explore hypothetical situations that require new appellations, such as *German citizen* (Justice Scalia = 2, Justice Stevens = 1) and *resident* (Justice Ginsburg = 1, Mr. Kneedler = 1). Both the Conservative Justice Rehnquist and the Justices who hold a liberal position, Justice O'Connell and Justice Stevens, use the term *client*, as do the two advocates defending the class of Mr. Zadvydas.

Justice Scalia has the greatest linguistic variety on the conservative side, saying *individual* ($n = 3$), *actor* ($n = 2$), *German citizen* ($n = 2$) and *felon* ($n = 1$). This is representative of his rich image-provoking vocabulary concerning the detention of criminal behaviour: "So even if you're talking about, you know, a real life Hannibal what's his name, you know, a really wicked, evil person who is going to harm people, there's every reason to believe that this person who's been deported because of serial murders, if you can't find a country to send him to, you have to let that person out?" (Justice Scalia appears to be referring to the blockbuster film *Hannibal* (2011), released in that same year, in which the main character is a cannibalistic serial killer named Hannibal Lecter.)

Concerning the liberal position discourse, Justice Ginsburg's affirmations tend to contrast with those of Justice Scalia, as her vocabulary includes 22 occurrences of expressions with *person/people* and five with an indefinite pronoun. She also uses *class* twice, plus there is one use each of *resident, shoplifter, infant* and *parent*. She (2001) draws on the ambiguity of *person*, which can apply to citizens or noncitizens alike: "if you are successful today, in any one of these situations, be it a shoplifter, be it someone who overstayed a visa and encountered a nasty INS person, that person could be locked up forever without any access to a judge, because the only thing is whatever process the administrator has chosen to give". Also, in this 2001 case, Justice Breyer is the only person among this selected sample to utter the appellation *human being* ($n = 1$).

In the *Jennings v. Rodriguez* (2017) case, expressions based on *person/people* ($n = 64$) surpass the quantity of occurrences of *alien* ($n = 54$). Almost half of these are credited to the advocate of Mr. Rodriguez's class, Mr. Arulanantham ($n = 28$). Only twelve of these occurrences are uttered by a conservative actor of the Court, Justice Roberts ($n = 4$) and the advocate of the government, Mr. Stewart ($n = 8$). In contrast, Justice Roberts and Mr. Arulanantham both use the term *outliers* ($n = 3$). The term *sub-class* ($n = 18$) is also common to Justice Sotomayor and both advocates. There is no other range of vocabulary within the conservative positions, but within the liberal position discourse we find *detainee* ($n = 8$), *arrivings* ($n = 4$), *client* ($n = 2$) and one each of *US citizen's wife, businessperson, immigrant, admits, noncitizens, sex offenders, parties* and *visitor*.

Finally, the 2020 *Johnson v. Guzman Chavez* case has the least lexical diversity. While in both 2001 and 2017, excluding forms of *alien* and *people/person* and indefinite pronouns, there are fourteen different terms to refer to noncitizens, but there are only six in 2020. Here, *alien* ($n = 72$) represents 52.2% of the total appellations. The term *individual* ($n = 21$) is the second most frequent, notably because of Mr. Hughes' ($n = 19$) liberal discursive practices, shared briefly by Justice Breyer ($n = 1$), and the opposing conservative Justice Gorsuch ($n = 1$). Other terms shared by both ideological sides include *noncitizen* ($n = 13$), *respondents* ($n = 5$), and *clients* ($n = 3$). Only the liberal Justice Sotomayor utters the remaining terms *immigrants* ($n = 2$), *applicants* ($n = 2$), and *class* ($n = 1$). Thus, despite the introduction of the appellation *noncitizen*, the conservative discourse remains dominated by the term *alien*. Furthermore, over the two decades of this selected sample of oral arguments related to detained noncitizens, the discussion of hypothetical or actual persons—and therefore their appellations—gradually diminishes.

5.3 Findings and discussion: ideology and absence of human presence

While quantitively examining the frequency of terminology within each ideological grouping, I noticed a recurrent attention to legal issues that resulted in a lack of references to human beings. This attention to legal questions is to be expected in certain court proceedings, such as in Example (3), where the qualitative nature of the absence of references to people becomes evident. In Example (3), Justice Thomas is questioning the advocate of Ms. Guzman Chavez, who was detained when she re-entered the country after having been previously ordered to leave.

Example 3): *Johnson v. Guzman* Chavez (2020)
Justice Thomas: Thank you, Mr. Chief Justice. Mr. Hughes, I'm a bit confused. Would you—so we had a removal order that **Respondent** was subject to. We agree on that. **Respondent** returns to the country. Now what happens to that underlying removal order in your assessment?
Mr. Hughes: Your Honor, for that underlying removal order to be reinstated, there has to be certain determinations that are made, which includes identifying that the **person at issue** is the same **person that was subject to and deported** on the first time pursuant to that removal order. Second, that the **individual**, in fact, unlawfully re-entered the country rather than—
Justice Thomas: Let's assume that's—let's assume those—that's accurate, that—that we're not debating the factual part. What's the effect of the process of reinstating removal?
Mr. Hughes: Your Honor, when you have that reinstatement order, it does bring back to life the underlying order of removal. And so, with the underlying order of removal and the reinstatement order paired together, that, when that process is final, does authorize the United States to remove an **individual** when that—when that process is concluded.
Justice Thomas: So what—why is that—how—how do you reconcile that? And this goes back to the Chief Justice's point. How do you reconcile that with what the Court said about CAT [Convention Against Torture] orders in Nasrallah?
Mr. Hughes: Two reasons, Your Honor. First is, when do we think that this process reaches conclusion? And we think the process reaches its conclusion, it's administratively final, at the point that it's—it's executable. And that's important because, otherwise, the categorical language of (a)(5) would seem to allow the government to remove an **individual** during the pendency of the withholding process. But the government doesn't take that position. My friend on the other side made that quite clear, that they cannot do so. That would be in violation of the statute. But my—my first point, if I can for a moment, is that we don't think the Court even needs to reach (a)(1)(B) and administrative finality, and that's because (a)(1)(A) is the gateway that defines what the removal period is. And if the removal period—if the government categorically does not have authority to remove, it makes no sense to conclude that the removal period, the 90 days during which the government shall remove, has begun. It simply doesn't meet the definition without even getting into the—the—(a)(1)(B).
Justice Thomas: Thank you.

Justice Thomas begins by referring twice to the *respondent*. *Respondent* may refer to the advocate or here, to the client. In the introduction to his question, Justice Thomas employs an inclusive *we*, which may refer to the participants of the oral arguments or to society in general, who observe the presence of a removal order. The second inclusive *we* focuses more particularly on the participants of the Court arguments, as there is consensus of agreement. This inclusive *we* contrasts with the lack of details concerning the respondent, who retains only an attachment to an imprecise *the county*.

Mr. Hughes refers to this specific class of noncitizens by reintroducing their humanity, through the use of *person at issue, person that was subject to and deported* and *individual*. Justice Thomas proposes to bypass the factual issues of the concerned person to focus on the effect of the process. He does so without referring to the effect on the given person of the process. In contrast, Mr. Hughes concludes that once the process is final, there is authorization to remove an *individual*.

Similarly, in Mr. Hughes' second answer to Justice Thomas, the advocate again highlights the place an individual of this given class plays in the process, as it "would seem to allow the government to remove an individual during the pendency". Contrary to Justice Thomas, Mr. Hughes also brings agency to the government. While Justice Thomas refers to the applicable legal aspects of procedure: "removal order", "process of reinstating removal" and the CAT, Mr. Hughes, considers that the government can "remove an individual", "take that position" or not, or "have authority". Thus, for Justice Thomas, the respondent remains vague and subject to actions while the focus is placed on the application of the law and the underlying procedures.

5.4 Findings and discussion: ideology and human presence

In a second example taken from the same 2020 case, Justice Sotomayor places the concerned class of people at the heart of her questions to Mr. Suri, who acted as defence for the conservative government. During the following exchange, Example (4), the central question is about the application and implications of two provisions, 1221–1226 and 1231 (a). The appellations range, although the legal class in question encompasses the specific category of persons detained for over three months.

Example 4): *Johnson v. Guzman Chavez* **(2020)**
Justice Sotomayor: Counsel, I'm not altogether sure I understood Justice Alito's question to you or your response. Can you clarify for me, how many **people** are held—are denied supervised release after the 90-day period?
Mr. Suri: After the 90-day period, I don't have exact statistics **on that**, but I took Justice Alito to be asking about the initial 90-day period, not after the 90-day period.
Justice Sotomayor: So, I'll ask about after the 90-day period. How many **people** are granted supervised release after that 90-day period?

Mr. Suri: I don't have the statistics on **that question**, but Respondent cites a study—cite [sic] a study that claims that **15 percent of the aliens** are released at some point before the withholding-only proceedings are completed.
Justice Sotomayor: So, **85 percent** are restrained?
Mr. Suri: That's correct.
Justice Sotomayor: Now 1231(a) commands that the agency release **immigrants**—remove **immigrants** within 90 days. **You** seem to have admitted to Justice Breyer that that, in most cases, is impossible for **this class of alien**. Now, if that is true, **we** have to pick between two provisions, 1221—1226 and 1231(a), and determine which controls the question of whether **these aliens** can be released on bond or bail or supervision. Why should **we** not adhere to the basic interpretive principle that counsels in favour of a harmonious reading and against choosing an interpretation that produces a substantial effect that is inconsistent with the text of another provision? If **we** accept **your** reading, basically, **we**'re saying for **this class of withholding-only—withholding applicants**, that **they**'re never going to be removed within 90 days of the finality of their removal order. That just doesn't make any sense to **me**.
Mr. Suri: I—I appreciate the force of the argument, but I have two responses to it. The first is that the obligation is subject by its own terms to the phrase "except as otherwise provided" in this section. Then, if **you** go down to 1231(a)(5), the last sentence of that states that **aliens with reinstated removal orders** can be removed "at any time." **We** think that "at any time" takes precedence over the 90 days. The second—
Justice Sotomayor: I'm sorry. I—I—I'm not quite sure I follow that argument. **You** already admitted to the Chief Justice that **you** can't remove these **people who are in withholding proceedings** because the law doesn't permit **you** to. [. . .]
Mr. Suri: I admitted to the Chief Justice that, as a practical matter, **these aliens** can't be removed if there's no country willing to accept **them**. That was the same situation, by the way, in Zadvydas, where the government was searching for a country to which to remove **the alien**. That was analysed under Section 1231.

Justice Sotomayor begins by placing her query in the wake of Justice Alito's question about the 90-day period. She evokes the factual question of how many *people are held* or *denied supervised release* after ninety days. Her use of the passive voice removes the presence of the government or other agencies who are acting to hold or to deny admission. The advocate, Mr. Suri, admits to a lack of information about *that*, replacing a reference to the number of people concerned with a common pronoun. This contrasts with his normative reference to Justice Alito.

Justice Sotomayor responds by placing herself first, *I*, as the central active agent. She then refers again to *people are granted*, using the passive voice that again avoids reference to the action of governmental agencies. Mr. Suri responds again with *that question*. His reference to the respondent, unlike that of Justice Thomas, refers to the opposing advocates who do have statistics concerning, as Mr. Suri says, the percentage of *aliens* who are released in the given context. Justice Sotomayor responds, repeating his data with an ellipsis—85 percent—avoiding a repetition of *alien*.

Justice Sotomayor's longer interrogation then unfolds with a series of different appellations. She begins with a reference to the legal text (1231 [a]) and reference to the actor (*the agency*), before referring this time to *immigrants*. She amends her verb choice from *release* to *remove* while again referring to the existence of *immigrants*. She then refers directly to Mr. Suri (*you*) and to his comment to Justice Breyer, before uttering a traditional reference, *this class of alien*. She echoes this refence with *these aliens* in the following sentence. Like Justice Thomas in Example 3, she uses an inclusive *we* to introduce the legal question of the possible provisions: "*we* have to pick between two provisions." Justice Thomas then asks why "should *we* not adhere to the basic interpretive principle" and wonders aloud what *we* are saying about the practical question at hand if *we* "accept your reading" of the provision. She continues her interrogation with a new reference, to *applicants*, before concluding with a new reference to herself, saying that it "doesn't make any sense to me"—thereby peopling her discourse on many levels.

Mr. Suri also begins by repeating the first person—*I appreciate* and *I have two responses*—but avoids direct reference to Justice Sotomayor by evoking *the argument* instead of y*our argument* and by saying *it* instead of *you*. For a third time, Mr. Suri utters the term *aliens*, qualifying it here "with reinstated removal orders". Although Mr. Suri is the sole advocate of the government, he concludes with *we think*. Thus, the exact parameters of these decisive actors, *we*, remain undefined. Justice Sotomayor interrupts, again placing herself as central agent, *I'm not quite sure* and *I follow*. Mr. Suri's undefined *we* is replaced with *that argument*, thereby erasing their human presence. Justice Sotomayor does, however, immediately refer to the singular *you* and to other members of the Court in the reminder: "you already admitted to the Chief Justice". In the same breath, Justice Sotomayor employs the pronoun *you* in a nebulously inclusive sense: "you can't remove these people" because the "law doesn't permit you to". She could have chosen an inclusive *we* by saying, *we can't remove* or *the law doesn't permit us to*. However, with the pronoun *you* she distances herself from the argument of Mr. Suri.

Mr. Suri also begins his response with the active subject *I* and with a reference to the Chief Justice. His answer repeats *aliens* twice: *these aliens* and *the alien*. In fact, both of these references are to stateless persons. No country is *willing* to accept *them*, because the people in question are not citizens of any country. The term *alien* covers a wide range of situations. In Mr. Suri's explanation, the term's imprecise generality erases the complex legal context of statelessness, one that goes beyond the motivation of a third country to *willingly* accept a given person. Mr. Zadvydas had spent over four decades in the United States as a *stateless person*—the depth of such experience is erased by Mr. Suri's appellation *the alien*.

6 Conclusions

This study has sought to understand oral discourse of actors of the US Supreme Court as it relates to detained noncitizens. The selected samples of three cases spans two decades and therefore a diversity of justices and advocates, whose discourse reflects and contributes to the national and international conversation on controversial questions of liberty of movement. The analysis of terminology related to the people concerned both by immigration and detention reveals a high correlation with ideological alignments. This correlation concerns both the specific terminology and touches upon the very subject of discussion, which may coldly exclude the human implications of a given law or procedure.

First, this study highlights the ideological preference by conservative advocates and, to a lesser degree, conservative justices, for the term *alien*. This term may be paired with *criminal*, in the case that a person has committed a crime unrelated to their administrative status. The adjective *illegal* is absent from the discourse of all parties. The liberal-leaning Mr. Stansell employs the adverb *illegally* when stating that the general rule is that "once an alien has passed through our gates, and it is more or less a direct quote, has passed through our gates, even *illegally*, they're entitled to the protections of the Fifth" (*Zadvydas v. Davis* 2001). Although Mr. Stansell notes that the passage "through our gates" may be *illegal*, he does not refer to a person as being *illegal*. The term *unlawful*, although infrequent, is preferred over *illegal*, irrespective of ideological stance. Thus, the appellation *illegal alien* is completely absent from the Court during these three cases directly related to detained noncitizens.

Second, many universal terms, such as *person*, *individual* or indefinite pronouns may be used to evoke citizens and noncitizens alike, thereby confirming the suspected dehumanizing implications of the term *alien*. These universal terms, as with *alien*, may be qualified in accord with a given administrative situation. Conservative Justice Rehnquist (2003) does so when he says that: "the Government says here that Mr. Kestutis is in the same position as someone, as an entering alien, who we've said has virtually no constitutional rights". Here, Justice Rehnquist begins with a more universal term, *someone*, before requalifying the person as *an alien*, highlighting the interchangeability of these terms, in spite of their contrasting ideological character.

Third, although the more recent term *noncitizen* is shared across ideological lines, liberal justices tend to exhibit greater linguistic variety. Introduced by a liberal advocate in the 2017 case, the appellation *noncitizen* continues to be used by a liberal advocate in 2020 and is notably adopted by the recent conservative justices, Justice Coney Barret and Justice Kavanaugh. Unlike the conservative justices, Justice Sotomayor employs it once amid many other appellations, such as *immigrant*, *applicants* and *aliens*. Overall, the increasing conservative attention to legal technicalities of the law, irrespective of their impact on the persons concerned, appears to reduce or eliminate the discussion of the persons concerned and of the corresponding terminology. Thus, as the more neutral term *noncitizen* comes into favour across ideological lines, conservatives are simultaneously paying less attention to the human implications of the application of the law, and are therefore using fewer and less varied appellations overall.

Finally, universal words such as *individuals, someone* and *people* contradict the notion of *us* and *them*—citizens or not. These linguistic borders of citizenship are also blurred when actors of the Court draw upon pronouns, such as *you* or *we*, which may refer to another member of the Court or to all of the members of the Court, the government or governmental agencies, immigrants, or even society at large. More research is needed to unravel the intricate play of pronouns within Supreme Court discourse to identify the power that these inclusive or exclusive words may play alongside ideologically influenced lexical choices.

References

Anthony M. Kennedy. Oyez. www.oyez.org/justices/anthony_m_kennedy (accessed 1 December 2022).
Baron, Jill & John DeSantis. 2017. Contested Subjects: Coalition-based Activism in the Library. *Association of College & Research Libraries/New England Chapter*, Burlington, VT, 12 May 2017.
Chemeriksny, Erwin. 2014. *The Case Against the Supreme Court*. New York: Penguin Books.
Cole, David. 2003. Their liberties, our security democracy and double standards. *International Journal of Legal Information* 31(2). 290–311.
European Parliament and of the Council of the European Union. 2013. Regulation of the European Union, No 603/2013 of the European Parliament and of the Council, 'Eurodac'. *Official Journal of the European Parliament*, 26 June 2013. https://www.easo.europa.eu (accessed 1 December 2022).
Evans, Robert J. E. 1999. Eurodac. Statement presented during the European Parliament Debate on Eurodac, Strasbourg, 18 November, 1999. https://www.europarl.europa.eu (accessed 1 December 2022).
Fox, Violet, Nick Bennyhoff, Kelsey George, Erin Grant, Tina Gross, Cate Kellett, Arden Kirkland, Karla Jurgemeyer, Treshani Perera, Karl Pettitt, Jeanne Piascik, Lisa Robinson & Deborah Tomaras. 2020. *Report of the SAC* [Subject Analysis Committee] *Working Group on Alternatives to* [*Library of Congress Subject Headings*] *(LCSH) 'Illegal aliens,'* Submitted to the American Library Association/ Association for Library Collections & Technical Service /Cataloging and Metadata Management Section Subject Analysis Committee, 19 June 2020.
Goldring, Luin. 2008. Minutes of Proceedings of the Standing Committee on Citizenship and Immigration Committee Meeting, House of Commons/Chambre des Communes, No. 028, 2nd Session, 39 Parliament of Canada, April 9, 2008. https://www.ourcommons.ca (accessed 10 December 2022).

Greenhouse, Linda. 2021. *Justice on the brink: The death of Ruth Bader Ginsburg, the rise of Amy Coney Barrett, and twelve months that transformed the Supreme Court*. New York: Random House.

Hartwell, Laura M. 2021a. Once an alien has passed through our gates: Noncitizens and their environments in three US Supreme Court oral arguments. *5th ILLA General Conference: Language and the Law in the Age of Migration*. Universidad de Alicante, Alicante, 7–9 September 2021.

Hartwell, Laura M. 2021b. Law, language and on-line translation tools. *The Triptych: National, European and International Law, The French way, International Association of Law Libraries Course*, Université de Toulouse Capitole, Toulouse, 4–9 October 2021.

Jennings v. Rodriguez. 2016. Oyez. https://www.oyez.org/cases (accessed 15 September 2022).

Kelsey George, Erin Grant, Cate Kellett & Karl Pettitt. 2021. A path for moving forward with local changes to the Library of Congress subject heading "illegal aliens". *Library Resources & Technical Services* 65(3). 84–95.

Laurence, Anthony. 2019. *AntConc*. https://www.laurenceanthony.net/software/antconc/ (accessed 15 March 2023).

Library of Congress, Policy and Standards Division. 2016. March 16. Executive Statement: Library of Congress to Cancel the Subject Heading "Illegal Aliens". https://www.loc.gov (accessed 15 September 2022).

New York City Commission on Human Rights. 2019. *New York City Commission on Human Rights Legal Enforcement Guidance on Discrimination on the Basis of Immigration Status and National Origin*. https://www1.nyc.gov/assets/cchr (accessed 20 September 2022).

Johnson v. Guzman Chavez. 2021. Oyez. https://www.oyez.org/cases/2020/19-897 (accessed 15 September 2022).

Rosen, Jeffrey. 2016. What made Antonin Scalia great. *The Atlantic*, 15 February 2016.

Sánchez, Linda. 2021. *United States House of Representatives 1177—U.S. Citizenship Act*. https://www.congress.gov/bill/117th-congress/house-bill/1177 (accessed 15 March 2023).

Sandra Day O'Connor. Oyez. www.oyez.org/justices/sandra_day_oconno (accessed 1 December 2022).

Scott, Ridley (dir.). 2011. *Hannibal* [film]. Metro-Goldwyn-Mayer.

Vitorino, António. 1999. *Eurodac*. Statement presented during the European Parliament Debate on Eurodac, Strasbourg, 18 November 1999.

Zadvydas v. Davis. 2001. Oyez. https://www.oyez.org/cases/2000/99-7791 (accessed 15 September 2022).

Mary C. Lavissière and Rachel C. Hill
Chapter 5
Improper entry by an alien on trial: the uncomfortable linguistic past of 8 United States Code § 1325 & 1326

1 Introduction

Between July 2017 and June 2018, thousands of minors were separated from their parents by American authorities while trying to cross the border into the United States. These family separations were permitted under the prosecution of parents for Improper Entry or Reentry by an Alien, 8 U.S.C. § 1325 & 1326, respectively. These statutes were originally codified in 1929, when hostile attitudes towards Latin American immigrants first led Congress to seek criminalization of unauthorized entry (Gonzalez O'Brien 2018). Several Supreme Court decisions question whether § 1325 & 1326 remain good law under the reasoning of the Supreme Court decision in *Village of Arlington Heights v. Metropolitan Housing* (1977), which holds that a law enacted with discriminatory motivation is unconstitutional. These statutes continue being challenged in a number of federal courts, including *United States v. Carrillo-Lopez* (2021) and *United States v. Machic-Xiap* (2021), as well as by researchers in legal fields (Fish 2021).

This paper contributes linguistic evidence to the current legal debate. Using linguistic methods, it demonstrates that these statutes were originally motivated by racial animus,[1] as shown from a corpus of legislative discourse surrounding their passage. It also contributes to legal linguistics by showing that the "language of the law" (Trosborg 1997) found in statutes cannot be fully separated from other contemporary legal and political discourses.

This paper has two objectives. First, we review early immigration legislature in the United States and bring attention to the evidence therein of racial animus. Specifically, we show that the records of the state legislative discourses at issue in *Ramos v. Louisiana* (2020) and *Hunter v. Underwood* (1985), as well as the record

---

[1] In *Arlington Heights* (1977) the Supreme Court of the United States defined racial animus as "negative intent or purpose with racist motivation".

Mary C. Lavissière, Nantes Université
Rachel C. Hill, esq. Legal consultant

pertaining to the original passage of Improper Entry by an Alien in 1929, all declare a racist belief that immigration poses a threat to "pure Americanism", as seen in Example (1) below.

> 1) Another reason why the quota should apply to the country south of the Rio Grande and the islands is because their population in the main is composed of mixed blood of white, Indian, and negro. This makes their blood a very great penalty upon the society which assimilates it. The United States already has sufficient race and blood troubles. (House 1928a: 2462)

Second, we add linguistic nuance to the dichotomy between the language of the law and other types of legal language or non-specialised language that surrounds it. We question the relevance of typologies of legal genres that separate "language of the law" from other types of legal language. The criteria used in *Arlington Heights* (1977) points to the necessary connection between the interpretation of legal statutes and other types of discourse surrounding their passage. This seems particularly relevant to the cases investigated here, as can be seen from statements made during the 1898 Louisiana Convention (Louisiana 1898) in Example (2).

> 2) It is a declaration upon the part of this Convention, that no white man in this State—*that's the effect but not the language*—that no white man in this State who has heretofore exercised the right of suffrage shall be deprived of it, whether or not he can read or write, or whether he possesses the property qualification. *That is the meaning of it: nothing more and nothing less.* It declares that every white man between now and the 1st day of September next, although he may not be able to read and write, although he does not possess the property qualification, may notwithstanding, if he register himself pursuant to this ordinance of the Constitution, be thereafter entitled to vote. *That is the meaning of it.* (Louisiana 1898: 375) (our emphasis)

In the above quotation, the members of this state constitutional convention, infamous for disenfranchising the Black citizens of Louisiana (Fish 2021: 52) through literacy tests and property ownership prerequisites for voting, made a clear distinction between what the language of the law would state and the meaning they gave the constitutional provisions, as well as being clear about the intended effect of the law. This distinction, overtly expressed in Louisiana, was later echoed by Representative O'Connor of New York in 1929, in his critique on the floor during national Congressional debate on the bills that would become 8 U.S.C. § 1325 & 1326, seen in Example (3).

> 3) [It] has been said around this Chamber that these *three bills are perfectly innocent* and do not involve much controversy. [. . .] I have always had grave doubt about any immigration bill which was presented to this House. *It is not so much the particular provisions of any immigration bill as it is the spirit behind the bill with which I am deeply concerned.* Here today I want to discuss that spirit which I believe pervades our country to an alarming

extent [. . .] I believe that there are many people in this country today who fear that these United States of ours is the most intolerant, narrow-minded nation on the globe. (House 1929b: 3526)

The similarity of the statements, even though made from opposing perspectives, demonstrates the importance of considering the discourse surrounding the language of the law in order to correctly interpret its meaning. The analysis in the present paper, therefore, contributes linguistic evidence to the current debate on efforts to strike down § 1325 & 1326 on the grounds that they were motivated by racial animus. (For a related discussion on the terms "racist motives" and "racist reasons" in the context of Spanish hate crimes see Chapter 9 in this volume.)

This paper is divided into four sections. First, in our literature review we review the case law that frames how racial animus has been defined and linked to laws in the United States. In this section, we also question literature that has separated language of the law, such as that found in statutes and contracts, from other types of legal language. Second, we describe the two corpora used for this study. Our reference corpus includes legislative and media discourse surrounding the passage of laws later struck down for having been motivated by racial animus. Our TEST corpus includes similar discourse surrounding the passage of 8 U.S.C. § 1325 & 1326. We also describe our methods, which include the textual analysis tools Iramuteq (Ratinaud 2014) and Sketch Engine (Kilgarriff et al. 2014). Third, we describe our results. Fourth, we discuss our results in the light of the relevant literature, and make recommendations about how the relationship between the language of the law and legal language might be viewed differently going forward.

2 Literature review

The following literature review is divided into two parts. First, we review the literature concerning the legal framework and history of 8 U.S.C. § 1325 & 1326. We include a history of the statutes themselves and a summary of case law that is relevant for the problem of determining whether or not a law is tainted by racial animus. Second, we review some of the work done on the typology of legal genres. Here, we focus on the relationship between the language of the law and legal language as identified by Trosborg (1997). We question Trosborg's separation of legally binding language, such as statutes and contracts, from other types of legal language such as legislative debate surrounding the passage of laws. Finally, we discuss the notion of genre. We integrate Bhatia's (2008) notion of "interdiscursivity" into the conversation concerning evaluation of racial animus motivation from a linguistic perspective. We also highlight the fluid nature of genres as discussed by Giltrow and Stein (2009).

2.1 Legal literature review

In the United States, it is a crime for a person without legal documentation to enter or to attempt to enter the country. Sections 1325 & 1326 enumerate both misdemeanour and felony variations of this crime. The maximum penalty for the felony offense is 20 years in prison. Today, immigration prosecutions account for the majority of federal criminal cases in the United States. Tens of thousands of people are charged and prosecuted with misdemeanour and felony immigration crimes every year (Fish 2021). Most recently, under the Trump administration, the misdemeanour immigration crime was used to prosecute parents of child immigrants, and to separate families at the border between Mexico and the United States.

These federal crimes were originally enacted in 1929, through the Undesirable Aliens Act. As the title of the law evinces, the debate surrounding the passage of this law included discussions of eugenics, the "mongrelisation" of US society, compelled by fears about miscegenation (Fish 2021). During the legislative discussion that took place, the proponents of the law presented a purported expert, Harry H. Laughlin, to support their acutely racialised view of immigration. They submitted Laughlin's reports into the record, stating their intent to "protect American blood from alien contamination" (United States & Laughlin 1928: 4). In particular, Laughlin was known for his model sterilization law.[2] Laughlin testified for the legislative record that "[i]mmigration control is the greatest instrument which the Federal Government can use in promoting race conservation of the Nation" (House 1928c: 19). Following these discussions and testimony, as documented in the Congressional Record, the Undesirable Aliens Act was passed. These immigration crimes have since remained a part of federal law in the United States, with little change or review to this day.

In US law, the purpose of the legal concept of "equal protection" is to ensure that laws are not discriminatory. The US Supreme Court has held that a law may violate this principle of equal protection in three ways. First, a law may discriminate on its face, such as a Virginia law criminalizing interracial marriage that was challenged in *Loving v. Virginia* (1967). Second, authorities may apply a law that appears

---

[2] In 1933 the Nazi government in Germany implemented "Gesetz zur Verhütung erbkranken Nachwuchses" (which translates "Law for the Prevention of Offspring with Hereditary Diseases") Teicher (2019: 289). This law bears striking similarity to Laughlin's model law released in 1928. Allen (1986: 252–254) discusses the relationship between Laughlin and the Nazi party and mentions the possible link between Laughlin's model law and the German one in note 72. Teicher (2019: 299–300) analyses the relationship between the measures of the two laws. He observes both similarities and differences, concluding that the legislators in Germany may have consulted Laughlin's law, though an exact copy was not adopted.

neutral on its face ("facially neutral"[3]) in a discriminatory way, a practice challenged in *Yick Wo v. Hopkins* (1886). In this case, a San Francisco ordinance made it illegal to operate a laundry in a wooden building without a permit. Only one permit was granted out of two hundred Chinese applicants, while only one among eighty non-Chinese applicants was denied a permit. And, third, a facially neutral law may be enacted with a discriminatory purpose to disparately impact a disfavoured group. This final form of discrimination was challenged in the case of *Village of Arlington Heights v. Metropolitan Housing Development Corp* (1977). *Arlington Heights* dealt with a zoning ordinance in Arlington Heights, a Chicago suburb, which barred the construction of multi-family housing facilities (such as apartment complexes). Ultimately, the Supreme Court upheld the law. While the court acknowledged concerns that the law's impact kept minorities and members of other historically disenfranchised communities from moving into the neighbourhood, it found that racial animus had not been a motivating factor in the law's passing.

This last method for an equal protection inquiry, articulated in *Arlington Heights* (1977), has provided an avenue for legal advocates and scholars to engage with the legislative records and the history of certain US laws, including laws which may appear neutral but which continue to serve an original, discriminatory intent. Indeed, litigation is currently proceeding in federal courts around the United States against immigration crimes in federal law, with advocates arguing that the original racist intent behind the creation of these laws, and their continuing disparate impact on Latin American people, violates the principle of equal protection.

In *Arlington Heights* (1977) the Supreme Court developed a four-factor balancing test to determine whether or not a law was enacted with racial animus. Racial animus is a negative intent or purpose with racist motivation. If racial animus can be demonstrated through balancing the following factors, then the law should be found unconstitutional as a violation of the Equal Protection Clause of the federal Constitution. The four factors are:

---

[3] A law is classified as facially neutral when the language of the law does not create a suspect classification or application, nor infringe on a fundamental right protected by the Constitution. In these instances, the court looks for a disparate impact to determine unconstitutionality. The distinction between the "face" of the law and its application was made by the Supreme Court in *Yick Wo v. Hopkins* (1886: 117–118): "Though the law itself be fair on its face and impartial in appearance, yet, if it is applied and administered by public authority with an evil eye and an unequal hand, so as practically to make unjust and illegal discriminations between persons in similar circumstances, material to their rights, the denial of equal justice is still within the prohibition of the Constitution." Later, in *Washington v. Davis* (1976: 426), the Court added the adjective "neutral" to the doctrine, "A statute, otherwise neutral on its face, must not be applied so as invidiously to discriminate on the basis of race."

1. the impact of the challenged decision (whether it bears more heavily on one race than another)
2. the historical background of the decision
3. the specific sequences of events leading up to the challenged decision
4. the legislative or administrative history, especially where there are contemporary statements made by members of the decision-making body

Typically, judicial deference is given to decision-making bodies, under the assumption that they have appropriately weighed many competing interests to arrive at a decision. The Supreme Court has stated that the test for violating the Equal Protection Clause does not require that the discriminatory intent was either dominant or primary. Instead, it states that: "when there is a proof that a discriminatory purpose has been a motivating factor in the decision, this judicial deference is no longer justified" (*Arlington Heights* 1977: 428–429).

The Supreme Court's more recent decisions in *Hunter v. Underwood* (1985) and *Ramos v. Louisiana* (2020), both of which relied on the *Arlington Heights* (1977) four-factor framework, are central to this legal discussion and to litigation surrounding the legislative history of federal immigration crimes. In *Hunter* (1985), the Supreme Court struck down a provision of the Alabama Constitution, which disenfranchised persons convicted of "any crime involving moral turpitude" (Alabama 1901: 530). In *Ramos* (2020), the Supreme Court struck down laws passed in Louisiana and Oregon, which permitted criminal convictions to be obtained from nonunanimous juries. Using the *Arlington Heights* (1977) four-factor balancing test, courts looked at the legislative history and the record of debate during the passage of these laws, to find that the "avowed purpose" of these laws was discriminatory.

The adoption of these Louisiana and Alabama constitutional provisions, as well as the Oregonian law at issue in *Ramos* (2020) and *Hunter* (1985), occurred at nearly the same time that federal immigration crimes were passed into law—the late 1920s and early 1930s. In addition, all the laws enacted were criminal laws. A linguistic comparison of the legislative record and other discourses—the stated purpose of these laws—provides evidence that, like the laws at issue in *Hunter* (1985) and *Ramos* (2020), the federal immigration crimes that were passed in 1929 have the same underlying discriminatory purpose.

2.1.1 Litigation of 8 U.S.C. § 1325 & 1326

Today, the courts are actively litigating the constitutionality of 8 U.S.C. § 1325 & 1326. In August 2021, a Nevada district judge declared § 1326 unconstitutional in *United States v. Carrillo-Lopez* (2021). Also, in *United States v. Machic-Xiap* (2021), an

Oregon district judge declared § 1326 constitutional while acknowledging its racist history and effects. These decisions are now on appeal at the Ninth Circuit Court of Appeals, and litigation may eventually reach the Supreme Court.

2.2 "Language of the law" versus "legal language"

A longstanding debate divides textualists from contextualists (Eskridge 1990: 621). This debate turns on the question: should the law be understood to be limited to the words on the page, the "plain meaning", or should it be understood within the context of the authors and passing of the legislation? Since the 1990s the Supreme Court has been shifting from a rebuttable presumption of the "plain meaning" of the being the same as the intent of the authors to a "new textualism". This new textualism emphasises the written words of a text over any contextual evidence (Eskridge 1990: 623). Even still, where the justices are attempting to understanding the motivation for passing a law rather the application of the law, contextual analysis remains an essential step in the legal analysis. The *Arlington Heights* (1977) framework in fact is a kind of law—case law—that requires thorough investigation into the context of the creation of the text of a given law.

This case law calls into question the separation between language with the force of law, and other types of legal discourse. Yet, some researchers working on legal linguistics have made a distinction between "language of the law", such as statutes and contracts with binding legal force, and "legal language" which encompasses a wider range of genres (Trosborg 1997). The latter includes exchanges in the courtroom, exchanges between lawyers and their clients, and other nonbinding genres pertinent to legal language. In this section, we call this typology into question. We find the typology unfit for situations in which US courts must determine the motivation behind the passage of a statute or the writing of a constitution, either at state or federal level, as a matter of case law.

Trosborg (1997) describes legislative legal writing, such as statutes and constitutions, as self-contained. She describes this type of legal writing as a communication process that is "complete and explicit". In her view, these characteristics separate it from other types of legal writing (1997: 28). She further claims that: "This process is one of 'intention-to-expression-to-interpretation' that is not to be found in other writing activities" (1997: 28). Trosborg represents statutes and constitutions as having "sincere intention, real meaning, and conforming action" (1997: 30).

In the light of case law such as *Arlington Heights* (1977), however, this description of the legislative process and statutory language omits an essential fact. Indeed, some legislators seek to hide the true intention of their legislation. This is clear from the statement of the President of Louisiana's 1898 Constitutional Convention "that's the effect but not the language" (Louisiana 1898: 375). At this convention, the legislative body of the State of Louisiana met to enact laws disenfranchising Black voters and to disempower Black jury members (Alabama 1901; Fish 2021). They sought to do so while evading sanctions from the federal government, which, through the Fifteenth Amendment of the US Constitution, had given Black men the right to vote. Their discriminatory intent was explicitly stated in the speeches given by lawmakers in the Louisiana legislature during the convention. The unmistakable contrast between the explicit intent of the legislators and the facially neutral language of the statute itself refutes Trosborg's (1997) description of legislative drafting.

A more accurate understanding of "language of the law" incorporates a family of genres and discourses from which it can never be completely extricated. Bhatia argues that it is necessary to look at "multiple discourses" in order to "develop a comprehensive and evidence-based awareness of the motives and intentions of disciplinary and professional practices" (2008: 165). The need to view a text as embedded in other discourses has been recognized by social scientists from diverse disciplinary backgrounds, who "suggest that the proper understanding of societies, social institutions, identities, and even cultures may be viewed as discursively constructed ensembles of texts" (Alvesson & Kärreman 2000: 137). Bhatia (2008) uses Alvesson and Kärreman (2000) to connect genre analysis to the concept of interdiscursivity. He defines this term as "appropriation of semiotic resources (which may include textual, semantic, socio-pragmatic, generic, and professional) across any two or more of these four levels, especially those of genre and professional practice" (Bhatia 2008: 165). Bhatia further emphasises that interdiscursivity is needed to understand what is explicitly stated as well as what is implicitly understood by a discourse community, to give expression to their "private intentions" within the socially accepted communicative practices and shared generic norms" (Bhatia 2012: 24).

Bhatia (2008) gives an example of interdiscursivity in the private sector in the case of Enron's annual reports. The collapse of the energy company is a particularly interesting study because it was unexpected and was followed by lawsuits related to accounting fraud. Enron's annual reports show evidence of a mixing of discourse types, the first of which is accounting discourse, the second of which is narrative. The latter is closer to public relations discourse. Narrative was used to prepare shareholders for a drop in Enron's performance. Thus, the narrative portions of the reports are used to understand the meaning and effect of the accounting portions. Bhatia's (2008, 2012) definition does not, however, fully resolve the problems raised by the discourses surrounding 8 U.S.C. § 1325 & 1326, because the statute and the legislative interactions adhere to the codes of their own respective genres. Similarly, the full meaning of a statute cannot be grasped without also analysing its legislative record. This was pointed out by Giltrow and Stein (2009: 4), who take a pragmatic approach to genres. In their analysis, genres come with their own "basis for inferencing". Citing new-rhetorical theories of genre, Giltrow and Stein (2009) argue that genres should be considered as historically contingent.

Seen from this perspective, the studies carried out both by Trosborg (1997) and Bhatia (2012: 2008) omit two important points for the study of 8 U.S.C. § 1325 & 1326. First, Trosborg (1997) describes statutes as a specific type of language, with no mention of their legislative history. In contrast, we argue that statutes are another type of institutional discourse and should also be viewed interdiscursively and situated socially and historically (Giltrow & Stein 2009). Second, and because of their social "situatedness", while some statutes are sometimes discriminatory on their face, it is equally possible that the motivating racial animus is not directly expressed in their language. As shown in examples 2 and 3, the drafters of this legislation rely on the "autonomy" of the legislative genes (Giltrow & Stein 2009: 6). They depend on the "calling-up on the pre-signal of a specific configuration of processing strategies, of types of schematic knowledge and inferencing strategies" (Giltrow & Stein 2009: 6) in their legislative peers and the general public. These strategies reduce complexity and focus cognitive processing. In this situation, only the discourses surrounding the passage, and not the words of the legislation itself, of these laws reveal their racist motivation and objectives. However, in contrast to the definition offered by Bhatia (2008), this type of interdiscursivity does not represent a hybridisation or bending of genres. The language of the statute itself is neutral on its face, while there is no apparent mixing of genres. For this reason, with Giltrow & Stein (2009), we argue that viewing genres as socially and historically situated is more appropriate. In other words, it is necessary to move outside of the language of the law *per se* to the language surrounding the passage of the law to understand the true intentions or motivations behind that law. In American Law, *Arlington Heights* (1977) provides a legal framework for understanding why the discourse that surrounds statutory language is in fact part of the law: it is incorporated into the institutional aspect of the law through case law.

In short, this paper makes three contributions to the conversation about language and law. First, we show that quantitative and qualitative linguistic analyses can contribute to the legal conversation about the similarity between the discourses produced during the passage of a law struck down for racial animus—as in *Hunter* (1985) and *Ramos* (2020)—and those surrounding the original passage of another statute suspected of the same motives (i.e. 8 U.S.C. § 1325 & 1326). Second, we show that the language of the law in US statutes cannot be represented as completely self-contained but is intimately connected to larger legal and nonlegal discourses surrounding its debate and ratification. Third, we argue that the definition of interdiscursivity must include situations which show no direct evidence of either the mixing or bending of genres.

3 Corpus and methodology

3.1 Corpus

To investigate these three points, *supra*, we collected two corpora. First, a reference corpus (hereafter REFERENCE) of the legislative record of the passage of the laws challenged in *Hunter* (1985) and *Ramos* (2020). This corpus is a reference corpus because these two cases focused on laws that were facially neutral but which were struck down by the Supreme Court as unconstitutional, based on finding that racial animus was a motivating factor in their passage. The decision to strike down these laws was determined after an analysis of the history of the law in the legislative records and other discourses surrounding the original passage of the laws. Second, we collected a test corpus (hereafter TEST) which includes legislative records from the passage of 8 U.S.C. § 1325 & 1326. The ongoing legal case involving these statutes makes the argument that 8 U.S.C. § 1325 & 1326 should be found to have been motivated by racial animus.

3.1.1 Reference

REFERENCE includes the legislative record for Article VIII, § 182, of the Alabama Constitution of 1901, struck down in *Hunter* (1985), and La. Const. art. I, § 17(A) and La.C.Cr.P. art. 782 A, struck down in *Ramos* (2020). Specifically, we include speeches from the proceedings of the constitutional conventions of Louisiana (Louisiana 1898) and Alabama (Alabama 1901). As stated in our literature review, these two conventions are infamous for their disenfranchisement of Black citizens (Fish 2021). The proceedings of these conventions are voluminous, and much of the discourse includes different genres of discourse such as: amendments proposed to the Constitutions, votes about the amendments, roll calls, minority reports, petitions received from external sources and read to committees, and speeches made to the entire convention. Because the amendments, votes, debates, and reports concern a wide range of legislative concerns, we chose to include only the speeches from the conventions which were related to the passage of the law in question. These highlight the main concerns of the legislators regarding the specific law at issue. The speeches take place at different moments in the convention. The details of these speeches can be found in Table 5.1.

In addition to the purely legislative discourse included in the cases of *Hunter* (1985) and *Ramos* (2020), REFERENCE includes press discourse that was cited as important for the passage of the nonunanimous jury laws in Oregon, a provision that was passed after a ballot vote by the Oregonian population (Kaplan & Saack 2017).

Table 5.1: REFERENCE corpus: speeches and number of words.

Convention	Convention Day (Date)	Speaker	Title (if available)	No. of Words
Alabama	1 21 May 1901	John B. Knox of Calhoun	Opening Statement	4,807
Alabama	82 3 Sep 1901	Chairman and District Representatives	Address of the Committee	7,366
Louisiana	1 3 Feb 1898	Hon. E. B. Kruttschnitt, Chairman of the Convention	Opening	2,131
Louisiana	6 14 Feb 1898	Dr. J. L. M. Curry, representing the Peabody Education Fund, and Ho. C. E. Fenner	Not Available	8,862
Louisiana	21 14 Mar 1898	Hon. William Jennings Bryan, of Lincoln, Nebraska	Science of Government	8,054
Louisiana	66 12 May 1898	Hon. Thomas J. Semmes, Chairman of the Committee on the Judiciary	Closing	9,834
Total Number of Words				33,000

As in Louisiana's case, Oregon's unanimous jury provision was also motivated by racial animus, specifically aimed at immigrants (Kaplan & Saack 2017). The details of the media sources can be found in Table 5.2.

Table 5.2: REFERENCE corpus: press articles included and number of words.

Journal	Date	Author (if available)	Title of article	No. of words
Oregon Taxpayer	Apr 1934	Not Available	Five Measures Call for Voters' Decision	959
Oregonian	26 Apr 1932	James Adam	Testimony Closes in Massie Trial	604
Oregonian	7 May 1932	Mr. Easton	"'Honor Case' Jury Upheld" [letter to the editor and reply]	344
Oregonian	3 Nov 1933	Not Available	Debauchery of Boston Juries	319
Oregonian	25 Nov 1933	Not Available	One Juror Against Eleven	379
Total Number of Words				2,605

3.1.2 Test

Our corpus TEST includes the legislative history and congressional debate on the floor of the US Congress for the passage of the 1929 law. The language of the 1929 law was re-enacted as 8 U.S.C. § 1325 & 1326 in 1952, with the only change being an expansion of policing power (*Carillo-Lopez* 2021: 5–6). An ongoing case questioning the constitutionality of § 1325 & 1326, *United States v. Rios-Montano* (2020), included exhibits of legislative records to demonstrate racial animus underlying *Arlington Heights* (1977). We use the exhibits from this case for our TEST corpus. The details of this legislative history can be found in Table 5.3 below.

Table 5.3: TEST corpus: speeches and reports included, sources and number of words.

Rios Montano Exhibit (Source)	Date	Chamber of Congress	Title of document (if available)	Genre	No. of words
B (House 1928c)	21 Feb 1928	House, Committee on Immigration and Naturalization	The Eugenical Aspects of Deportation	Discussion	20,134
C (House 1929a)	16 Feb 1929	Committee of the Whole House on the State of the Union	Deportation Of Aliens	Discussion	12,047
D (House 1924)	5 Apr 1924	House	Not Available	Speech	23,012
E (House 1928b)	19 Jan 1928	Immigration Conference	Restriction of Mexican Immigration	Speech	2,788
F (House 1928a)	27 Jan 1928	House	Immigration and the Crime Wave	Speech	2,381
G (House 1926)	12 Jan 1926	Committee on Immigration and Naturalization	Deportation	Discussion	15,844

Table 5.3 (continued)

Rios Montano Exhibit (Source)	Date	Chamber of Congress	Title of document (if available)	Genre	No. of words
H (Senate 1929b)	17 Jan 1929	Committee on Immigration	Making it a Felony with Penalty for Certain Aliens to Enter the United States of America Under Certain Conditions in Violation of Law	Report	757
I (Senate 1929a)	23 Jan 1929	Senate	Not Available	Discussion	897
J (House 1929c)	6 Feb 1929	Committed to the Committee of the Whole House on the State of the Union, From the Committee on Immigration	Deportation Of Aliens	Report	4,977
K (House 1929b)	15 Feb 1929	Committee on Immigration and Naturalization	Immigration	Discussion	15,379
Total Number of Words					98,216

For the Iramuteq analysis, both corpora were coded with the variables CORPUS, DOCUMENT, DATE, and GENRE, as seen in Table 5.4 below.

Table 5.4: coded variables of REFERENCE and TEST corpora.

Variable	Definition	Category	Definition	Example
CORPUS	The corpus in which the document is included	reference	Document is part of reference corpus of laws judged as motivated by racial animus according to legislative proceedings.	*corpus_reference
		test	Document is part of test corpus and is an exhibit in the case United States of America v. Bruno Rios-Montano.	*corpus_test
DOCUMENT	The title of the document or exhibit	Various	An abbreviated form of the document's title	*doc_al_const, *doc_exhibit_b
DATE	Document's publication date	Various	Date in numbers	*date_1901
GENRE	The form given to the information in a document, and which corresponds to a function in a given speech community	speech	A unit of discourse that addresses the legislative body as a whole by a leader of the body.	*genre_speech
		press	Short articles from local newspapers in Oregon exhorting a change in laws about unanimous jury verdicts.	*genre_press
		discussion	Legislative proceedings concerning a bill, an amendment, a hearing, or a report requested by the legislative body.	*genre_discusssion
		report	A report submitted by a member of the legislative body about which there is no discussion in the proceedings.	*genre_report

3.2 Methodology

This paper uses two software programs for quantitative analysis, Iramuteq (Ratinaud 2014) and Sketch Engine (Kilgarriff et al. 2014). We also include qualitative analyses based on Bhatia's theories about genre, as outlined in the literature review (Bhatia 2008, 2012; Bhatia & Salmani-Nodoushan 2015). We describe the basic functions of the software programs in the paragraphs below.

3.2.1 Iramuteq

Iramuteq (Ratinaud 2014) is a software containing algorithms with a geometric approach to data analysis (Loubère & Ratinaud 2014; Ratinaud & Marchand 2012). Many of the algorithms are based on the pioneering work on statistical analyses of textual data carried out by Jean-Paul Benzécri and his students (Beaudouin 2016; Greenacre 1984). The algorithm used in this paper is a clustering algorithm that creates statistically significant classes of words that appear frequently, arranged in a descending hierarchical classification (DHC) (Reinert 1983; Ratinaud & Marchand 2012; Schonhardt-Bailey et al. 2012). While the software creates the classes, it does not name or interpret them. Interpretation is the work of researchers and scholars. In this article, we name and interpret the classes generated by Iramuteq.

Before the algorithms are applied, Iramuteq divides the corpus into text segments (TS). We chose a segment-length of 45 words, which corresponds with the average length of sentences in legal texts (Hiltunen 2012). The corpus is then lemmatised and divided into three types of *forms*. *Active forms* include adjectives, adverbs, nouns, and verbs. *Supplementary forms* include grammatical words such as conjunctions, prepositions, articles, etc. *Hapax legomena* (hereafter *hapax*) include forms that only appear once in the corpus. All algorithms are applied to the lemmatised corpus.

3.2.2 Sketch Engine

Sketch Engine is a corpus query and data management software that allows for comparison between corpora as well as between typical language usage and usage anomalies. While the DHC integrated into Iramuteq allows us to see the larger lexical structure of discourse, Sketch Engine (Kilgarriff et al. 2014) shows how lexical units behave in terms of grammar and collocations. In this way, the more localised lexical focus of Sketch Engine complements Iramuteq's comprehensive representation. We use the features Word Sketch and Thesaurus offered by Sketch Engine (Kilgarriff et al. 2014).

4 Results

4.1 Results from Iramuteq

4.1.1 DHC of REFERENCE

The DHC of REFERENCE yields 7 classes, as can be seen in Figure 5.1. Below, we describe these classes and provide examples of representative text segments.

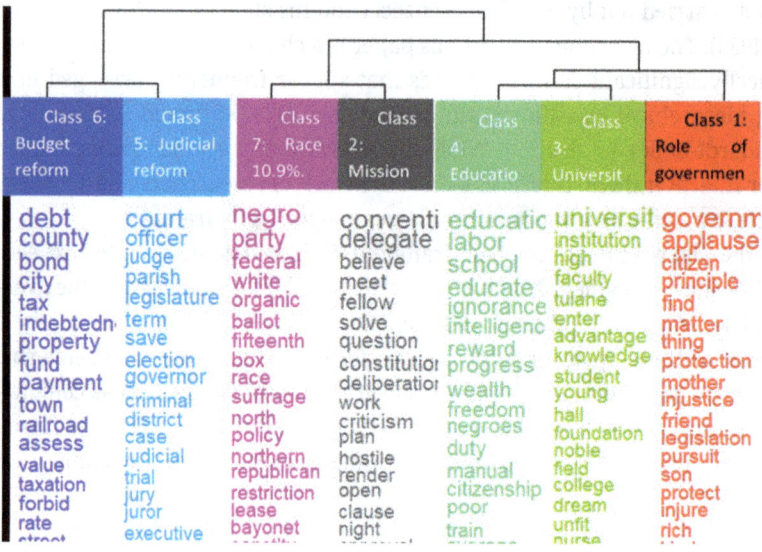

Figure 5.1: Classes resulting from the dhc of reference.

Class 1, *Role of government*, represents the role and values attributed to government in its relation to citizens, as in Example (4).

> 4) But a *people* cannot always *live* in a state of revolution. The time comes when, if they would be a free, *happy* and *contented people* they must return to a constitutional *form* of *government*, where *law* and order prevail, and where every *citizen stands* ready to stake his life and his honour to maintain it. (Alabama 1901: 13)

Class 2, *Mission*, represents the description of the mission of the convention as presented by the leading members of the legislative community at various points of the conventions. An example of a statistically significant extract is provided in Example (5).

> In the first place, my *fellow*-citizens, we are all aware that this *convention* has been called by the *people* of the *State* of *Louisiana* principally to deal with one *question*, and we *know* that but for the existence of that one *question* this *assemblage* would not be *sitting* here to-*day*. (Louisiana 1898: 9)

This class contains forms that would be expected in opening and closing addresses to a legislative body assembled to amend a constitution, such as *fellow-citizens, convention*, and *assemblage*. In addition, however, this example refers to the central issue for the convention in terms of *that one question*. The speaker goes as far as to state that that question—covertly, the vote of new freemen and women—is the sole reason for the convention.

Classes 3 and 4, which are called *University* and *Education* respectively, contain the educational issues treated by the legislative bodies. Examples (6) and (7) below are typical of these classes.

> 6) [. . .] in that *great* constitution an amendment embodied in the act for the incorporation of *Tulane University* the *great object* of which is provided for in the preamble to *develop* and *establish* a *great university* in this State—the State at last *laid* the *foundation* for *accomplishing* the *object* which has haunted its *dreams* since the date of its corporate *existence*. (Louisiana 1898: 38)

> 7) The *negroes* will remain a constituent portion of *Southern population* and *citizenship*. What are to be our *relations* to them? Are they to be *lifted* up or *left* in the *condition* of discontent, *ignorance*, poverty, semi-barbarism? (Louisiana 1898: 32)

Class 5, *Judicial reform*, contains segments of discourse related to the state judiciary branches of Louisiana and Alabama. Examples (8) and (9) below are typical of this class. The reforms related to these systems, among other changes, opened more elected positions in Alabama and reduced the number of jurors on grand juries in Louisiana. In Louisiana, the reform also allowed for nonunanimous verdicts. TS (text sections) from the group of press articles related to the nonunanimous jury law in Oregon also appear in this class.

> 8) You have *saved jail fees* because of the fact that the *judges* will be able to *clear* the *jails* by quicker *trials*. I have heard it stated that the total *expenses* of the *judicial system* was stated by the gentleman who *preceded* me $330,000 per *annum*. (Louisiana 1898: 382)

> 9) The *Judicial system* has been so molded as to better meet *local* wants and more elastic *power* given to *abolish* or consolidate constitutional *courts*, when circumstances warrant the *executive department* has been strengthened and the *legislative department* improved in many ways *heretofore* pointed out. (Alabama 1901: 1775)

In these examples, changes are explicitly linked to budget reduction or the efficiency of the judicial system, although they are also linked to discrimination against minorities (Kaplan & Saack 2017).

Class 6, *Budget reform*, gathers TS related to decreasing state debt through changes in expenditures and taxes related to private companies and municipalities. Examples (10) and (11) are characteristic of this class:

> 10) *Cities, towns* and villages, with a few *excepted* by name whose necessities and *debts* demanded the *exception*, are *forbidden* to *levy* a higher *rate* of *taxation* per *annum* than *one half* of 1 per *centum* upon the *property*, as *assessed* for *state taxation* during the preceding *year*. (Alabama 1901: 1762)

> 11) They took appeals to the Supreme Court of the United *States*, and the *tax* laws of the several *States* which I have mentioned have been *maintained*, the result being that instead of driving a *revenue based* upon the *mere actual* market *value* of the *property* if put up at auction [. . .] (Louisiana 1898: 377)

Class 7, the final class, is named *Race* as it contains TS that explicitly mention race-related concepts. Among these we observe claims of the inherent supremacy of whites and inferiority of non-whites, as seen in the Examples (12) and (13) below.

> 12) There is a *difference* it is *claimed* with great *force*, between the uneducated *white* man and the ignorant *negro*. There is in the *white* man an inherited capacity for government which is wholly wanting in the *negro*. (Louisiana 1898: 382)

> 13) Our mission was, in the first place, to *establish* the *supremacy* of the *white race* in this *State* to the *extent* to which it could be legally and constitutionally done, and what has our *ordinance* on *suffrage*, the constitutional *means* by which we *propose* to maintain that ascendency, done? We have *established* throughout the State *white* manhood *suffrage*. (Louisiana 1898: 375)

In addition to the explicit goal of installing *white supremacy* through *white manhood suffrage*, the speakers highlight the lengths to which the conventions went to circumvent laws preventing discrimination.

The class *Race* and its close relationship with *Mission*, underline the extent to which racial animus motivated the constitutional conventions of Louisiana and Alabama. In this way, the discourse structures, as represented by the DHC, are similar between TEST and REFERENCE. In REFERENCE, the classes *Race* and *Mission* are closely related. In TEST, the classes *Immigration* and *Eugenics* are also related. These similarities call into question the extent to which certain forms are specific to each corpus. We study this question in the following section.

4.1.2 DHC of TEST

The following paragraphs describe the results from the TEST corpus of exhibits drawn from the legislative history of 8 U.S.C. § 1325 & 1326. We describe the results from the DHC. A list of example sections of text is provided for each of the classes shown in Figure 5.2. In each example section of text, the words which appear in the class are in italics. While we have named the classes, the figure is otherwise automatically generated by the software.

First the DHC results in four classes, as seen in Figure 5.2.

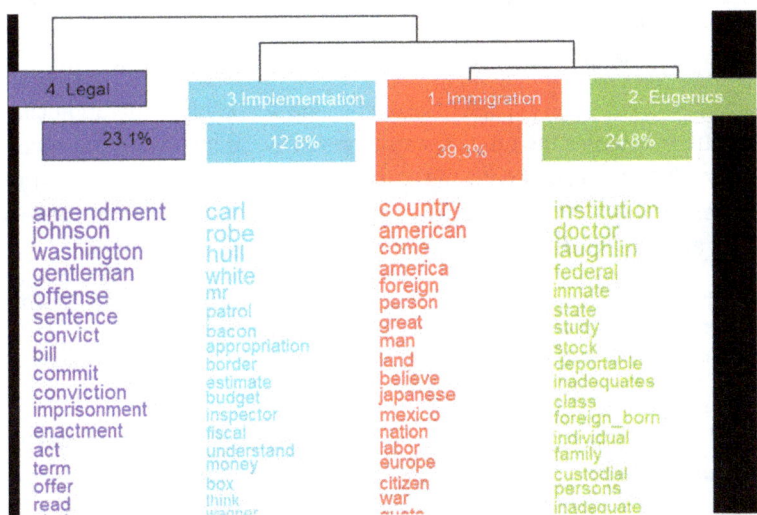

Figure 5.2: Classes calculated by the DHC of test.

Class 1 represents the largest percentage of the forms. We name it *immigration*, as it expresses the attitude towards immigration represented in the discourses surrounding the passage of 8 U.S.C. § 1325 & 1326. Examples of characteristic segments (according to the chi-squared test) can be found in Example (1) (cited in our introduction) and (14) below. Italicised words in each example are words that are included in the class by the program.

> 14) I do not *tell* you that these things will *come* to *pass*, but I do say we already have enough *Japanese, Chinese, Italians, Negroes* and other *foreign* strains. It is time to entirely stop the *islands, Europe, Asia* and Africa from dumping their *scum* and riffraff on our beautiful *American shores*. (House 1926: 2462)

These examples clearly show a hostile attitude towards immigrants coming from certain countries. Specifically, the use of words such as "scum", "riffraff", or "mixed blood", and statements that it is a "great penalty" for the country receiving these immigrants, are of a kind with the pejorative representations of immigrants from Latin American countries. These examples raise concerns related to assimilation of foreign cultures. For a related discussion on the misconceptions of memory, language adaptation, and trauma see Chapter 3 in this volume.

Class 2 represents the second largest percentage of forms. We name it *eugenics*, as it mainly contains statements by Laughlin (United States & Laughlin 1928). Example (15) is characteristic of this class.

> 15) Up to that *date* three *reports* had been *printed*. On that *date Doctor Laughlin* made a *preliminary report* on the subject "The *Eugenical Aspects* of *Deportation*" and he was *authorized* by the committee to *prepare tables* and analyses concerning *biological* and other *matters derived* from studies of the deportations from the United States. (House 1928c: 1)

Class 3 is the smallest class and represents the *implementation* of immigration control in the corpus. It includes the numbers associated with immigration and finances. These are mainly associated with the United States Border Patrol, as seen in Example (16) below. It is also interesting to note the reference to a literacy test, a device that was also used to disenfranchise Blacks in Louisiana and Alabama.

> 16) *Mr.* FREE. Is the *literacy* test being applied strictly along the *Mexican border* now? *Mr.* ROBE CARL WHITE. Much more so than ever before in the history of immigration. That is one of the reasons for the great decrease in the *number* of *Mexicans* coming in *legally* through our *ports*. (House 1926: 25)

Class 4 is called *legal* because it represents legislative discourse, such as protocols for proposing amendments or taking the floor from another member of Congress. It also includes forms related to the legal framework, such as sentencing, which were put into place to criminalise entry or re-entry of undocumented noncitizens into the United States. Example (17) is statistically representative of this class.

> 17) *Mr. JOHNSON* of Washington. Mr. *Chairman*, I *offer* the following *committee amendment*: The *clerk read* as follows:
> *Page* 4, after *line* 9, insert a new *subsection* to read as follows:
> (8) An *alien* who is *convicted* of carrying on or about the person *transporting*.
> (House 1929b: 3547)

4.2 Results from Sketch Engine

The algorithms provided in Sketch Engine allow us to make three major findings. First, courts often hold that immigration and domestic discrimination are separate legal issues requiring different considerations. However, our corpora provide evidence that legislative proceedings at the end of the 19th century involving domestic criminal laws, and those that took place in the 1920s–30s in federal Congress, aimed at immigration laws, were both deeply motivated by beliefs about race. This can be seen in Example (1) (cited in our introduction). In Example (18), below, the speaker states that individuals coming from diverse races would cause *degradation* of the US gene pool and that any white ancestor of these individuals is a *sorry white*.

> 18) This blend of low-grade Spaniard, *peonized* Indian, and negro slave mixes with negroes, *mulatoes*, and other *mongrels*, and some *sorry whites*, already here. The prevention of such *mongrelization* and the *degradation* it causes is one of the purposes of our laws which the admission of these people will tend to defeat. (House 1928b: 2817–18)

Second, our corpora show that lawmakers repeatedly used pejorative language to refer to individuals of non-white or racially mixed heritage. In Example (18), this language includes *peonized, mongrels*, and *mongrelization*. Third, that these beliefs motivated the proposal of the bills and their enactments. For instance, the speaker in Example (18) explicitly states that the laws concerning the criminalization of immigration of undocumented individuals from Latin America were centrally motivated by beliefs about race. The use of Sketch Engine allows us to show that these findings are not limited to Example (18) but, rather, reoccur in the discourse surrounding the passage of the bills that would become 8 U.S.C. § 1325 & 1326. In the following paragraphs, we therefore describe collocations and grammar surrounding several words related to race and immigration found in the corpus.

In Figures 5.3a and 5.3b, we compare the collocations of the word *race* in both corpora. There are several similarities in how this word is used in the corpora. With nouns modified by *race*, both corpora represent problematic racial relationships (see Chapter 2 in this volume for a discussion of the concept of danger in social relations and its link to criminality) through the collocation *racial conflict*, as in Examples (19) and (20).

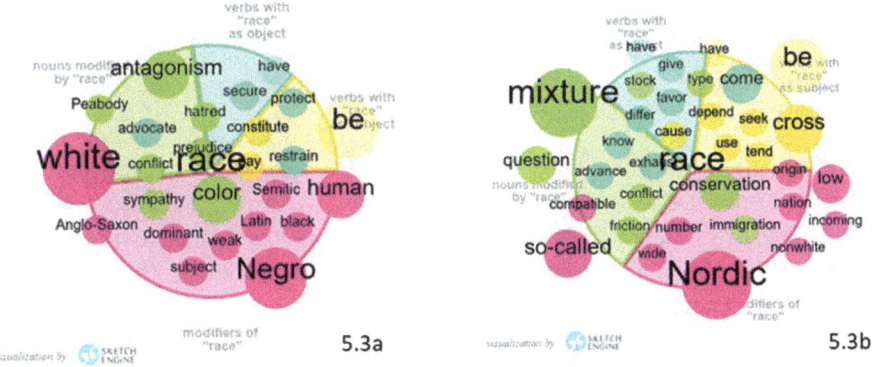

Figures 5.3a and 5.3b: Collocations of race in REFERENCE and in TEST.

> 19) The Southern people, with this grave problem of the races to deal with, are face to face with a new epoch in Constitution-making, the difficulties of which are great, but which, if solved wisely, may bring rest and peace and happiness. If otherwise, it may leave us and our posterity continuously involved in *race conflicts*, or, what may be worse, subjected permanently to the baneful influences of the political conditions now prevailing in the State. (Alabama 1901: 9)

> 20) Every incoming race causes blood mixture, but if this were not true, a mixture of blocs of people of different races has a bad effect upon citizenship, creating more *race conflicts* and weakening national character. This is worse when the newcomers have different and lower social and political ideals. (House 1928b: 2818)

Crucially, it is not only the use of the same collocation in REFERENCE and TEST, but also in both examples the use of an if-conditional to imply that if measures are not taken, racial conflict will be a constant and permanent threat. In (19), we see this threat represented through *continuously involved* and *subjected permanently*. In (20), the specter of further conflict is evoked through *creating more race conflicts*. The permanent nature of the effects are evoked through *weakening national character*. Both corpora also show legislators predicting a more dire situation in the future through *what may be worse* (19) and *this is worse when* (20). Other references to tension between races are also observed in both corpora; REFERENCE includes the expressions *racial prejudice*, *racial antagonism* and *racial hatred*. TEST has the term *racial friction*.

The modifiers of *race* in both corpora also have similarities. REFERENCE includes a larger variety of words referring to individual races: *white*, *Anglo-Saxon*, *negro*, *black*, *Latin*, *Semitic*, whereas in TEST a dichotomy between *Nordic* and *nonwhite* individuals is marked.

> 21) which we have reared in order to protect the purity of the ballot box, and to perpetuate the supremacy of the *Anglo-Saxon race* in Louisiana. If they do, theirs be the responsibility for bringing back the methods which have prevailed in the past. (Louisiana 1898: 380)

> 22) It may surprise some Members of this House to know that neither the second largest number of those deported, nor the third largest number of those deported, came from any of the much-maligned southeastern European countries, but from the so-called *Nordic races*, so much championed here. (House 1929b: 3545)

Interestingly, while the speaker in REFERENCE uses *Anglo-Saxon race* seriously, the speaker in TEST uses *Nordic races* with distance and sarcasm. He places *so-called* before the expression and cites statistics to counter the *much-championed* argument that certain geographical regions are home to superior individuals. This evinces that the debates about immigration in Congress were closely linked to notions of race.

The idea that certain ethnic groups are better than others is also present in both corpora. Both REFERENCE (23) and TEST (24) refer to certain ethnic groups with the expression *lower race*.

23) How can you adjust the relations of a higher and a *lower race*, each with the same rights and privileges while the government, Democratic as to one, cannot be safely made democratic as regards the other? (Louisiana 1898: 32)

24) The women of the lower classes and the so-called inferior races tend to take mates, whether legitimately or illegitimately, from the dominant or upper races. The consequence is that the perpetuity of a race depends upon the virtue of its women, and among the *lower races*, wherever two races come in contact, there is a tendency towards "breeding up" by the "pure sire method". (House 1928c: 19)

An examination of the views expressed about race in Congress during the criminalization of undocumented immigration is completed by observing the collocations with *blood* and *race* found in TEST. These collocations reveal the strong influence that the eugenics movement had on legislative decisions about immigration and domestic laws about marriage during the 1920s and 1930s in the United States (Fish 2021). See Figure 5.4 and Examples (25), (26), and (27).

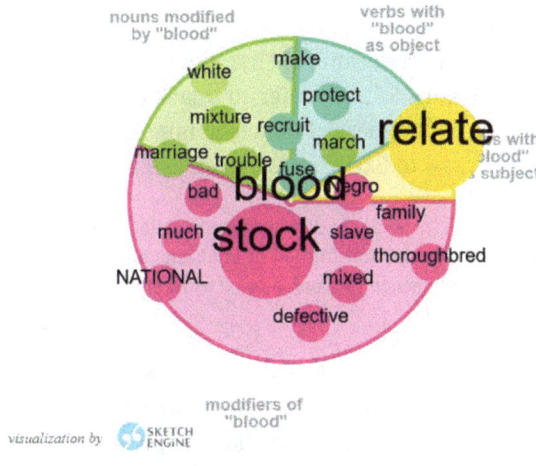

Figure 5.4: Collocations of *blood* in test.

25) [Would] it not be reasonable to believe that we would assume a secondary place as compared to those nations which have kept their *blood* white and purely Caucasian? I do not tell you that these things will come to pass, but I do say we already have enough. (House 1926: 2462)

26) [Excluding] the individual without reference to the stock or the family he comes from is not very effective, because in the next generation, instead of the alien feeble minded, for instance, supplying only 30 per cent of their quota, the children of immigrants supplied 190 per cent which showed the letting in of bad *blood*. The time will come when this country will have to

face, more courageously than it has at the present time, the matter not only of race and of individual quality, but also of pedigree or family stock. (House 1928c: 20)

27) Under the perfect operation of the law there would be no one to deport, but the fact is that many inadequates and potential inadequates have broken through our first lines, so that our last resort is to deport them, if we wish to protect American *blood* from alien contamination. (House 1928c: 4)

In addition to these examples, we can add Example (1) (cited in our introduction). Examples (25), (26), and (27), as well as Example (1), illustrate that those promoting the criminalization of immigration in TEST imply that it is better to *keep blood white* and *purely Caucasian* (25), than that there is *bad blood* (26). *Mixed blood* is represented as a *great penalty* (1) and *contamination* (27).

In addition to words referring to ethnicity and race, we also find words referring to specific groups of people. These include *negro* in REFERENCE and *alien, immigrant,* and *Mexican* in TEST. These words are used to refer to the individuals towards whom the racial animus of the legislative bodies is directed. The collocations are represented in Figures 5.5a–5.5d below.

We observe several similarities in the collocations. First, both include *incompetent* (Figure 5.5a) and *inadequate* (Figure 5.5b) to refer to a lack of skills. This presumed lack of productive capacity in members of domestic minorities and individuals of foreign origin is clearly linked in Example (28) from REFERENCE.

28) The exception in the Massachusetts law was no doubt directed against illiterate and *incompetent* immigrants, whereas, the provisions in the Constitution of Louisiana and North Carolina were directed against illiterate and *incompetent* negroes, as well as foreigners. (Alabama 1901: 16)

Other similar expressions in TEST include *dependent alien* (Figure 5.5b) and *degenerate immigrant* (Figure 5c), which are used by promoters of the criminalization of undocumented entry or re-entry. In contrast, those opposed to the law use *poor immigrant* or *helpless immigrant* to represent immigrants as nonthreatening.[4]

29) [. . .] influences who have no intelligent information of conditions, but who have a fixed obsession on Anglo-Saxon superiority, who have an obsession as to religious dominance, and who believe that it is proper to take vengeance upon them humble, harmless, *helpless immigrants*, in the course of the work allocated by themselves to themselves, and in so doing believe they are rendering service to their country. I feel sorry for them. (House 1924: 5887)

---

4 See also Chapter 4 in this volume for a consideration of formal appellations of US noncitizens by actors of the United States Supreme Court.

Figures 5.5a, 5.5b, 5.5c, 5.5d: Collocations of *negro* in REFERENCE and *alien, immigrant,* and *Mexican* in TEST.

It is also of interest to note *Jewish* in strong collocation with *immigrant* and *immigration* with *Mexican*. In 1921, a quota had been placed on Jewish immigration to the USA while Mexico extended an open invitation to Jewish immigration, leading a significant number of Jewish asylum seekers to emigrate to Mexico in the 1920s.[5] Thus, a compounding of racism is demonstrated here.

Finally, we compare the collocations of *negro* (Figure 5.5a) with *Mexican* (Figure 5.5d). One important observation is that both collocations include reference to an economic role as *laborer*, as seen in Example (30) from REFERENCE and Example (31) from TEST.

---

5 https://www.worldjewishcongress.org/en/about/communities/MX (accessed 25 October 2022).

30) I doubt not they would be met, in spirit, as the *negro laborers* from the South were met at the State line of Illinois, with bayonets, led by a Republican Governor, and firmly but emphatically informed that no quarter would be shown them in that territory. (Alabama 1901: 9)

31) The importers of such *Mexican laborers* as go to farms at all want them to increase farm production, not by the labour of American farmers, for the sustenance of families and the support of American farm life, but by serf labour working mainly for absentee landlords on millions of acres of semiarid lands. (House 1928b: 2817)

Labour represented by individuals immigrating from Latin America is represented as a threat to *American* [US] *farmers*. Labour provided by Latin Americans is so closely associated with the identity of the persons that the two are treated as effectively synonymous. That the speakers presumed this to be their whole identity is evinced by the use of the expressions *serf labour* in Example (31) and *Mexican peon labourer* in Example (32) from TEST.

32) Cheap labour today may be an expensive liability tomorrow, and surely in the *Mexican peon labourer* this condition has obtained. The American labourer resides among us, pays taxes, contributes to the welfare and upbuilding of society and really stands at the helm of the ship of state of our mighty Nation. On the contrary, foreign labour drifts into our country, obtains what it can for its hire, gives as little return as possible, in most cases, and then invariably thrusts itself for a charitable existence upon society, in the founding of which it has not assisted. (House 1928a: 2462)

The representation of Latin Americans as lazy and dependent on charity, with the incongruous "exception" of their labour capacity, echoes the representation of formerly enslaved Blacks who returned to the South after meeting hardship in the North (Example [33] from REFERENCE).

33) "Why, you damn black rascal, what are you stopping here for? Go into the kitchen and tell the cook to give you something to eat". "Before God, master", the old darkey said, grinning from ear to ear, "them's the sweetest words I'se heard since I left old Dixie". The old man was home at last. He was among people who understood him, and whom he understood. (Alabama 1901: 12)

In sum, REFERENCE and TEST respectively represent Blacks and Latin Americans as subjugated sources of labour, through words such as *master* or *serf*. These two corpora represent these groups of individuals as unable to hold power and responsibility, but rather as dependent for their sustenance on the benevolence and charity of dominant social groups. This representation is furthered by the pejorative language used to refer to these groups, words such as *peon* (Example [32]) and *darky* (Example [33]).

5 Discussion

In this paper, we have aimed to make three contributions to the literature about language and law. First, we brought evidence of the linguistic similarity between the legislative exchanges around *Hunter v. Underwood* (1985) and *Ramos v. Louisiana* (2020), and those surrounding the original passage of 8 U.S.C. § 1325 & 1326. The similarity exists at a corpora-wide level and a lexico-grammatical level. On a corpus-wide level, analyses such as the DHC show that TEST and REFERENCE contain similar discourse classes. Both corpora show that race and ethnicity are represented as problems to be solved through newly drafted legal instruments: new state constitutions in REFERENCE and new statutes in TEST. The legal measures proposed are presented as necessary because of the inherent inferiority of one group (non-whites) of individuals as opposed to another (whites). The latter is represented as superior and under threat of cultural and genetic contamination. The instruments are therefore specifically motivated and justified by a belief that one group should have less access and less power than another.

On a lexico-grammatical level, analyses of collocations in both corpora reinforce our claim that the corpora are similar. Both corpora show that the legislators involved used negative adjectives or pejorative language towards the groups most affected by the new legal instruments. Specifically, both immigrants and Blacks are represented as intellectually inferior, morally incapable of holding power, and dependent on the economic charity of others and therefore, dangerous. Both groups are equally presented as economic, cultural, and genetic threats to the dominant group. As such, each legislation is a measure enacted to protect the "purity" of the "vote" (REFERENCE) or of the "blood" (TEST). These findings show how linguistic tools can be useful in the process of evaluating the intent or motivation of laws that may not be legible in the facially neutral wording of said laws. If challenges on the basis of racial animus are to be brought forward in US jurisdictions, linguistic tools will be needed. Such challenges are made possible in part by the framework established in *Village of Arlington Heights v. Metropolitan Housing Development Corp* (1977), which includes discourse as one of the four factors relevant in determining whether or not a law was enacted with racial animus.

Second, both from a legal perspective, through *Arlington Heights* (1977), and from a linguistic perspective, we question Trosborg's (1997) typology of legal language. In US jurisdictions, the language of the law in statutes and state constitutions cannot be represented as isolated from other genres of legal discourse. This is in line with Giltrow and Stein's (2009) analysis of genre as socially-situated. Despite the "efficiency" (Giltrow & Stein 2009: 7) of statutes and state constitutions in the United States, legislative records, press articles, statements and interviews from legislators sponsoring, lobbying for, or drafting a legal provision must also be seen as contributing the meaning of the statutes and state constitutions themselves. As evinced by the explicit statements of legislators in the Alabama and Louisiana conventions, statutes and state constitutions have sometimes been purposely drafted in a facially neutral manner in order to avoid being struck down for discrimination. Our evidence challenges the textualist approach to law. We find that the social and legal meaning of a text cannot be complete when the text is isolated from the discourses surrounding its creation and implementation. In sum, the drafting process of documents that Trosborg claims represent the language of the law is more complex than "intention-to-expression-to-interpretation" (Trosborg 1997: 28). Trosborg further claims that "legislative expression forms part of a cycle of communication which, unlike most other kinds of written discourse, is complete and explicit" (1997: 28). In contrast, in line with Giltrow and Stein (2009), we find that legislative drafting must be analysed as part of a larger set of discourses. The text of the statutes themselves cannot be separated from the discourses that surround their drafting, particularly when assessing the original motivation for the legislation.

Third, we show that the current terminology concerning interdiscursivity cannot be limited to the hybridisation of statutes. The analytical framework must account for the embeddedness of one genre within another, even if no evidence for this connection is observed in the language of the statutes. In contrast, the use of the term interdiscursivity applies to a situation in which one genre colonizes another, such as in the case of Enron's annual reports (Bhatia 2008). Bhatia's (2008) analysis shows how traits from one genre can appear in another. In the case of the state constitutions and statutes analysed in this article, however, the legislators and drafters deliberately excluded language that would reveal the true objectives of these legal instruments. The true meaning of the legal instruments only becomes apparent when the discourses emitted during their drafting are read. In this way, Bhatia's (2008) definition of interdiscursivity is not helpful in describing the relationships between the texts studied here.

A possible alternative would be Bazerman's (1994) concept of *system of genres*, defined as "interrelated genres that interact with each other in specific settings" (82). Bazerman argues that a system of genres is

> the full set of genres that instantiate the participation of all the parties [. . .] the full interaction, the full event, the set of social relations as it has been enacted. It embodies the full history of speech events as intertextual occurrences but attending to the way that all the intertext is instantiated in generic form establishing the current act in relation to prior acts. (1994: 83)

A closer look at Bazerman's description of the systems of genres, however, reveals that the author envisions a tight chronological development of genres. The production of one genre closely precedes and conditions the next genre. However, the relationship between the discourses surrounding the passage of the statutes and the language of the statutes themselves is less clearly coordinated. This is especially true when a court is considering the original motivations behind the statute many decades after its passage, as is the case of 8 U.S.C. § 1325 & 1326. Rather than speaking of a system of genres or a chaining of genres, it seems more appropriate to speak of a *network of genres*, with dynamic links between different types of documents that are formed in ways that are not always systematic and can be reinterpreted temporally.

6 Conclusion

In sum, our research in linguistics informs not only literature about genre and discourse analysis, but may also contribute to discussions about interpretation in the legal field. While this is a complex question in law, current issues surrounding discrimination and immigration in the United States may benefit from linguistic analyses of the discourse surrounding the passage of the original statutes or constitutional provisions. Presenting the linguistic reality of statutes and other legal instruments as embedded and unalienable from their surrounding discourses may bring new nuance to debates about legal meaning and interpretation.

A possible extension of this research will be to further investigate the similarity in the representations of immigrants to the United States and racial minorities holding citizenship in the United States. While these two categories of individuals have so far been dealt with as separate cases in US law, the linguistic evidence suggests that similar negative representations and pejorative language are used to refer to these two categories of individuals. In this light, representations of both immigrants and US ethnic minorities as unworthy of access to the same rights as other groups are clearly present in the discourse of the legislators. The close relationship between these representations may help legal researchers and professionals gain a more unified view of legal questions related to civil rights of US minorities and immigration.

One limitation of our research is its firm anchoring in jurisdictions subject to US law. While we continue to claim that statutes, regardless of their jurisdiction, should be interpreted with an awareness of original intent (as manifested in discourses surrounding them), the legal weight of these surrounding discourses varies according to each legal system. In particular, *Arlington Heights* (1977), which gives legal weight to discourse surrounding the original passage of statutes, is only binding on jurisdictions subject to US law. In addition, and in connection with *Arlington Heights* (1977), the history of race and law in the United States has taken a particular trajectory that may or may not be relevant to other cultural contexts. Our research may, therefore, shed light on novel issues at the interface between law and language, but the particularities of the legal and cultural context of each jurisdiction will need consideration. Despite these limits, we hope our study will inspire linguists and legal professionals to collaborate in their endeavours to interpret the intentions behind textual genres invested with binding legal force, through investigations into the linguistic traits of discourses surrounding their creation.

References

Adam, James. 1932. Testimony Closes in Massie Trial. *Oregonian*, 26 April 1932.
Alabama. 1901. Journal of the proceedings of the Constitutional convention of the state of Alabama, held in the city of Montgomery, commencing May 21st, 1901. Montgomery: The Brown Printing Company.
AL Const. 1901. Art. VIII § 182.
Allen, Garland E. 1986. The Eugenics Record Office at Cold Spring Harbor, 1910–1940: An essay in institutional history. *Osiris* 2(1). 225–264.
Alvesson, Mats & Dan Kärreman. 2000. Taking the linguistic turn in organizational research: Challenges, responses, consequences. *The Journal of Applied Behavioral Science* 36(2). 136–158.
Bazerman, Charles. 1994. Systems of genres and the enactment of social intentions. In Aviva Freedman & Peter Medway (eds.), *Genre and the New Rhetoric*, 79–101. London: Routledge.
Beaudouin, Valérie. 2016. Statistical analysis of textual data: Benzécri and the French School of data analysis. *Glottometrics* 33. 56–72.
Bhatia, Vijay K. & Mohammad Ali Salmani-Nodoushan. 2015. Genre analysis: The state of the art (an online interview with Vijay Kumar Bhatia). *International Journal of Language Studies* 9(2). 121–130.
Bhatia, Vijay K. 2008. Genre analysis, ESP and professional practice. *English for Specific Purposes* 27(2). 161–174.
Bhatia, Vijay K. 2012. Critical reflections on genre analysis. *Ibérica* 24. 17–28.
Easton, T.S. & Oregonian editors. 1932. "'Honor Case' Jury Upheld" [letter to the editor and reply]. *Oregonian*, 7 May 1932.
Eskridge, William N. Jr. 1990. "The New Textualism". UCLA Law Review 37. 621–691.
Fish, Eric S. 2021. Race, history, and immigration crimes. *SSRN Scholarly Paper*. Rochester, NY: Social Science Research Network. https://papers.ssrn.com (accessed 15 March 2023).
Giltrow, Janet & Dieter Stein. 2009. *Genres in the Internet: Issues in the Theory of Genre*. John Benjamins Publishing.
Gonzalez O'Brien, Benjamin. 2018. *Handcuffs and chain link: Criminalizing the undocumented in America*. Charlottesville, VA: University of Virginia Press.
Greenacre, Michael. 1984. Theory and applications of correspondence analysis. The Food and Agriculture Organization (FAO). London: Academic Press.
Hiltunen, Risto. 2012. The grammar and structure of legal texts. In Peter M. Tiersma and Lawrence M. Solan (eds.), *The Oxford Handbook of Language and Law*, 39–51. Oxford: Oxford University Press.
House. 1924. Commercial Interests, 5887–5901. United States: Library of Congress. https://www.congress.gov (accessed 20 January 2022).
House. 1926. Deportation: Hearings before the Committee on Immigration and Naturalization, House of Representatives, Sixty-ninth Congress, first session, January 12, 1926. [. . .] proposed Deportation Act of 1926. Washington: G.P.O. https://catalog.hathitrust.org (accessed 20 January 2022).

House. 1928a. Immigration and the Crime Wave, 2461–2462. United States: Library of Congress. https://www.congress.gov (accessed 20 January 2022).
House. 1928b. Restriction of Mexican Immigration, 2817–2818. United States: Library of Congress. https://www.congress.gov (accessed 20 January 2022).
House. 1928c. The eugenical aspects of deportation: Hearings before the Committee on Immigration and Naturalization, House of Representatives, Seventieth Congress, first session, February 21, 1928, including testimony taken April 28, 1926. Washington: U.S. Govt. Print. Off. https://catalog.hathitrust.org (accessed 20 January, 2022).
House. 1929a. Deportation of Aliens, 3614–3621. United States: Library of Congress. https://www.congress.gov (accessed 20 January 2022).
House. 1929b. Immigration, 3525–3550. United States: Library of Congress. https://www.congress.gov (accessed 20 January 2022).
House. 1929c. Report No. 2397: Deportation of Aliens. United States: Congress.
Hunter v. Underwood, 471 U.S. 222 (1985).
Kaplan, Aliza B. & Amy Saack. 2017. Overturning *Apodaca v. Oregon* should be easy: Nonunanimous jury verdicts in criminal cases undermine the credibility of our justice system. *Oregon Law Review* 95(1). 1–52.
Kilgarriff, Adam, Vít Baisa, Jan Bušta, Miloš Jakubíček, Vojtěch Kovář, Jan Michelfeit, Pavel Rychlý & Vít Suchomel. 2014. The Sketch Engine: Ten years on. *Lexicography* 1(1). 7–36.
Loubère, Lucie & Pierre Ratinaud. 2014. Documentation IRaMuTeQ 0.6 alpha 3. http://www.iramuteq.org (accessed 22 February 2023).
La. Const. art. I, § 17(A).
La.C.Cr.P. art. 782 A.
Louisiana. 1898. Official journal of the proceedings of the Constitutional convention of the state of Louisiana, held in New Orleans, Tuesday, February 8, 1898. And calendar. New Orleans: Printed by H. J. Hearsey. https://catalog.hathitrust.org (accessed 20 January 2022).
Loving v. Virginia, 388 U.S. 1 (1967).
Ramos v. Louisiana, 140 S.Ct. 1390 (2020).
Ratinaud, Pierre & Pascal Marchand. 2012. Application de la méthode ALCESTE à de "gros" corpus et stabilité des "mondes lexicaux": Analyse du "CableGate" avec IRaMuTeQ. *Actes des 11eme Journées internationales d'Analyse statistique des Données Textuelles*, 835–844.
Ratinaud, Pierre. 2014. IRaMuTeQ: Interface de R pour les analyses multidimensionnelles de textes et de questionnaires. Windows, GNU/Linux, Mac OS X. http://www.iramuteq.org (accessed 20 January 2022).
Reinert, Max. 1983. Une méthode de classification descendante hiérarchique: Application à l'analyse lexicale par contexte. *Les cahiers de l'analyse des données* 8(2). 187–198.
Schonhardt-Bailey, Cheryl, Edward Yager & Saadi Lahlou. 2012. Yes, Ronald Reagan's Rhetoric Was Unique—But Statistically, How Unique? *Presidential Studies Quarterly* 42(3). 482–513.
Senate. 1929a. Regulation of Immigration, 2092. United States: Library of Congress. https://www.congress.gov/bound-congressional-record (accessed 20 January 2022).
Senate. 1929b. Report No. 1456: Making it a Felony with Penalty for Certain Aliens to Enter the United States under Certain Conditions in Violation of Law. United States: Congress.
Teicher, Amir. (2019). Why did the Nazis sterilize the blind? Genetics and the shaping of the sterilization law of 1933. *Central European History* 52(2). 289–309.
Trosborg, Anna. 1997. *Rhetorical strategies in legal language: Discourse analysis of statutes and contracts*. Tübingen: Gunter Narr Verlag.
Undesirable Alien Act. S. 5094, 70th Cong (1929).

United States v. Carrillo-Lopez, 3:20-cr-00026-MMD-WGC (D. Nev. Aug. 18, 2021).
United States v. Machic-Xiap, 3:19-cr-407-SI (D. Or. Aug. 3, 2021).
United States v. Rios-Montano, Case No.: 19-CR-2123-GPC, 2 (S.D. Cal. Dec. 7, 2020).
Unknown. 1933a. Debauchery of Boston Juries. *Oregonian*, 3 November 1933.
Unknown. 1933b. One Juror Against Eleven. *Oregonian*, 25 November 1933.
Unknown. 1934. Five measures call for voters' decision. *Oregon Taxpayer* 3(4), 1 Apr 1934.
Village of Arlington Heights v. Metropolitan Housing Development Corp., 429 U.S. 252 (1977).
Washington v. Davis, 426 U.S. 229 (1976).
Yick Wo v. Hopkins, 118 U.S. 356 (1886).

Jūratė Ruzaitė
Chapter 6
Impoliteness categories in hateful online comments targeting migrants in Lithuania

1 Introduction

This paper investigates pragmatic mechanisms and impoliteness strategies in hateful online comments directed against migrants on Lithuanian news portals, with the aim of outlining some potential classifiers of hate speech. Digital public domains have become toxic with verbal attacks on socially vulnerable groups, and migrants are one of the main groups targeted in online hate speech (cf. Chapter 8 in this volume). To counteract hate speech (legally and/or through social activism), numerous attempts have been made to develop stringent criteria for (automated) hate speech identification. However, there is still no agreement as to which features constitute the most optimal classifiers, nor in which language(s) these classifiers are most effective for identifying hate speech. The first software developed to identify offensive online comments in Lithuanian[1] (Petkevičius & Ruzaitė 2018–2020) also needs to be further refined for identifying hate speech, as it cannot currently distinguish between merely offensive content and legally sanctionable hate speech.

The problem of hate speech has been addressed from a variety of perspectives, including discourse analysis, corpus/computational linguistics and pragmatics. From the latter perspective, there has been a recent upsurge in researchers' concern with impoliteness strategies, such as mock politeness (Taylor 2015), name-calling (Vasilaki 2014), slurs (Bolinger 2015; Croom 2015), irony as an impoliteness tool (Tselika 2015), and the general phenomenon of impoliteness proposed by Culpeper (2011a, 2011b). These categories and models have proven effective when dealing with impolite behaviour in general and, as the current chapter aims to show, can be used when studying the abusive content of hateful online comments.

To the best of my knowledge, the only attempt so far to explicitly relate hate speech with linguistic impoliteness is Culpeper, Iganski and Sweiry's (2017) study. By examining the language manifested as religiously aggravated hate crime, Culpeper, Iganski and Sweiry address the broader question of whether the impoliteness model

---

[1] Neapykantos/įžeidžios kalbos socialinių medijų tekstuose atpažintuvas [Software to identify hateful/offensive content in social media]. http://hatespeech.vdu.lt/ (accessed 20 September 2022).

Jūratė Ruzaitė, Vytautas Magnus University

https://doi.org/10.1515/9783110789157-006

can be applied in the analysis of the language of hate crime, and which linguistic characteristics can be considered indicators of religiously aggravated speech (as a type of hate speech) (Culpeper et al. 2017: 2–3). In their analysis, they examine the most frequent types of impoliteness formulae and explore how they are intensified in hate speech. Interconnected with their work is that of Haugh (2015) and Haugh and Sinkeviciute (2019), who relate impoliteness with offence. Haugh and Sinkeviciute, however, do not account for hate speech in general or classifiers of hate speech in particular.

General principles of Impoliteness and Speech Act Theory are applied by Carney (2014) to analyse a court's interpretation of the Equality Act in a hate speech case. However, Carney's study also undertakes the more general task of assessing to what degree linguistic-pragmatic analysis can be applied in forensic linguistics, but does not account for impoliteness formulae in a systematic manner nor resort to a larger amount of data.

Another relevant, but less closely related, research focus in this area is the work on (im)politeness in relation to emotions (Langlotz & Locher 2017), as well as work on disagreement in computer-mediated communication (e.g. Shum & Lee 2013). If hate speech is considered a form of extreme disagreement, and if hate speech is assumed to involve strong emotions, some connections can be observed between the present work and these other studies, even though they make no specific reference to hate speech. However, these connections are only tentative, and of a very general nature.

Therefore, to address the lack of research directly relating hate speech and impoliteness strategies, the present study examines conventional forms of hostility addressed against migrants in online comments, by way of the following research questions:
1) What impoliteness formulae are used in Lithuanian anti-migrant comments online?
2) How much flexibility and creativity do these conventional structures permit? What structures remain relatively stable and can be used for automated identification of hateful content?
3) If only conventional impoliteness strategies are considered, what remains unaccounted for? What discursive strategies should be added to the impoliteness model (if any) when analysing online hate speech?

These research questions are answered on the basis of data obtained from the comments section on one major Lithuanian news portal, Delfi.[2]

---

2 https://www.delfi.lt/ (accessed 20 August 2022).

Section 2 below provides an overview of (im)politeness theory, and argues for its relevance in understanding hate speech. Section 3 is a detailed account of the data used in my empirical analysis, while Section 4 discusses the findings of the current research.

2 Impoliteness and hate speech

As Culpeper (2011b) and Grainger (2011: 167–188) outline, there are three "waves" in (im)politeness research.[3] The first wave stems from early theories in linguistic pragmatics, and focuses on the micro-level of interactions, deals with Speech Act Theory and conversational implicature, and favours a quantitative dimension in its research (e.g. Lakoff 1977). The second wave integrates the sociocultural dimensions of (im)politeness strategies, takes a discursive approach and emphasises broader societal dimensions. An important shift occurred in 2008 (Bousfield 2008), when impoliteness started being viewed as "strategic, systematic, sophisticated and not uncommon" (Culpeper & Hardaker 2017: 206) instead of being conceptualised as "some kind of politeness failure" (Culpeper & Hardaker 2017: 207).

Until the end of the 20th century, politeness theories predominantly focused on how communicative strategies are employed "to maintain or promote social harmony" (Culpeper 1996: 349). This approach thus accounted mainly for cooperative communication, following the work of such scholars as Leech (1983), Brown and Levinson (1987) and Lakoff (1989). Extended models, which included hostile communication alongside cooperative, did also exist (e.g. Craig et al. 1986; Lakoff 1989; Penman 1990). However, these models did not account for impoliteness in a more systematic manner, nor did they develop a comprehensive model that would cover specific strategies of impoliteness, the conventionality of such strategies, or the deliberate use of such strategies to express hostility. In the mid-1990s and later, impoliteness began to be addressed comprehensively by Culpeper (1996, 2005, 2009, 2010, 2011a, 2011b, 2013) and Culpeper, Bousfield and Wichmann (2003), who conceptualised it as "the use of strategies that are designed to have the opposite effect [of promoting social harmony]—that of social disruption" (Culpeper 1996: 350). In this model, impoliteness includes interactive strategies aimed at attacking face, and those deliberately leading to conflictive communication. Leech initially maintained that conflictive communication is "marginal to human linguistic behaviour in normal circumstances" (1983: 105), while Culpeper (1996 and in later works) goes on to demonstrate quite the opposite.

---

[3] For a more detailed review, see Culpeper and Hardaker (2017).

In the latter model of impoliteness, the concept of face, adopted from Goffman (1967: 5), is of special importance. As Culpeper, Iganski and Sweiry note, according to Goffman's definition, face "is not just the positive values that you yourself want, but what you can claim about yourself from what *others* assume about your actions" (2017: 7; [emphasis in the original]). Since the speaker's hostility can intentionally demean the victim's identity, they further stress the interconnectedness between face and identity when arguing the following:

> From the point of view of face, hostile insults are not about the psychological properties of either the victim or the defendant before the event in question, but about assumptions concerning what might conflict with the kind of positive values (including those involving identity) the victim might want to claim during a particular interaction. (Culpeper et al. 2017: 7)

Thus, in research on hostility and impoliteness, emphasis is laid on the fact that "face involves the public mediation of identity" (Culpeper et al. 2017: 7).

The relationship between impoliteness and hate speech lies in factors such as (1) intentionality/deliberate behaviour, (2) socially inappropriate ways of expressing/articulating verbal aggression, and (3) face-threatening behaviour, often resulting in emotional consequences, "such as humiliation or embarrassment" (Culpeper 2013: 1). However, as noted by Culpeper (2013: 1), research on the emotional impact of impoliteness remains limited and insufficient for making any broad generalisations.

Regarding speaker intent, intent has been addressed as a key characteristic in some (but not all) approaches to impoliteness (e.g. Bousfield 2008: 72; see also Culpeper 2013: 201–225; Culpeper & Hardaker 2017). Culpeper's (2005: 38) early definition of impoliteness refers to intent in a very definite manner:

> Impoliteness comes about when: (1) the speaker communicates face-attack intentionally, or (2) the hearer perceives behaviour as intentionally face-attacking, or a combination of (1) and (2). (Culpeper 2005: 38)

In this definition, both the speaker's intent, and the hearer's perceptions of intent, are emphasised.

Although Bousfield (2008) also stresses intent as a central feature of impoliteness, he suggests a more moderate perspective by proposing an approach based on four prototypical (but not obligatory) features of impoliteness:
(1) speaker intent/projectability
(2) speaker awareness of possible face-damaging effects of their utterance(s)
(3) hearer perception/construction of the speaker's intent/hurtfulness of their words,leading to
(4) hearer face actually being, or not being, damaged (Bousfield 2008: 124)

In Bousfield's model, prototypical cases of impoliteness do involve an element of intent, including the speaker's awareness of their own intent and the hearer's perception of that intent. But Bousfield does not view intent as a necessary characteristic of impoliteness, and some instances of impoliteness may even be inadvertent. In some later publications, Culpeper (2010: 3233) and Culpeper and Hardaker (2017: 203–204) also stress that not all impoliteness is intentional, and that full intent is not a necessary condition of impoliteness.

Culpeper (2010, 2011b) elaborates his earlier definition, and deemphasises intent by stating the following:

> Impoliteness is a negative attitude towards specific behaviours occurring in *specific contexts*. It is sustained by expectations, desires and/or beliefs about social organisation, including, in particular, how one person's or group's identities are mediated by others in interaction. *Situated behaviours* are viewed negatively when they conflict with how one expects them to be, how one wants them to be and/or how one thinks they ought to be. Such behaviours always have or are presumed to have *emotional consequences* for at least one participant, that is, they *cause or are presumed to cause offence*. Various factors can exacerbate how offensive an impolite behaviour is taken to be, including for example whether one understands a behaviour to be strongly *intentional or not* (Culpeper 2010: 3233; [my emphasis]).

As Culpeper and Hardaker (2017: 204) admit, this definition is very broad (especially in contrast to the earlier one), but it still usefully points to the central features of impoliteness, such as: situatedness, (not necessarily conscious) intent, face-damaging/harmful effects, and how social norms and expectations influence a hearer's perception of, and the speaker's construction of, an utterance. All these features are characteristic, at least to some extent, of hate speech, which is also usually defined as context-dependent/situated intentional behaviour, realised through aggressive verbal forms and targeted at certain groups of people protected by law.

To account for impolite behaviour, Culpeper (1996: 356–357) proposes five impoliteness superstrategies:

(1) *Bald on record impoliteness* – the FTA [face-threatening act] is performed in a direct, clear, unambiguous and concise way in circumstances where face is not irrelevant or minimised.
(2) *Positive impoliteness* – the use of strategies designed to damage the addressee's positive face wants.
(3) *Negative impoliteness* – the use of strategies designed to damage the addressee's negative face wants.
(4) *Sarcasm or mock politeness* – the FTA is performed with the use of politeness strategies that are obviously insincere, and thus remain surface realisations.
(5) *Withhold politeness* – the absence of politeness work where it would be expected (Culpeper 1996: 356–357).

Culpeper's superstrategies invert the politeness superstrategies proposed by Brown and Levinson (1987: 69), but, as Culpeper (1996: 356–357) stresses, it is important to distinguish *bald on record* impoliteness in his model from Brown and Levinson's (1987). For Brown and Levinson (1987), bald on record is a *politeness* strategy applied in specific circumstances. For example, in cases when face concerns are disregarded in an emergency, the threat to the hearer's face is minimal (e.g. "Come in" or "Do sit down"), or when the speaker is more powerful than the hearer (e.g. "Stop complaining" said by a parent to a child). In such cases, the face threat is minimal, and, crucially, the speaker does not intend to attack the hearer's face (Culpeper 1996: 356–357).

Another major contribution to the theory of impoliteness is the categorisation of impoliteness strategies and conventionalised formulae proposed by Culpeper (2010), based on Terkourafi's (cf. 2002) work on formulaic politeness expressions in Greek. With this framework, Culpeper provides a systematic linguistic repertoire of impoliteness forms, which include the categories listed in Table 6.1 (square brackets are designed to give an indication of some of the structural characteristics of the formulae; slashes are used to indicate alternatives).

Table 6.1: Impoliteness formulae in English (based on Culpeper 2010: 3242–3243).

Category	Formulae
Insults	
1. Personalised negative vocatives	– [you] [fucking/rotten/dirty/fat/little/etc.] [moron/fuck/plonker/ dickhead/berk/pig/shit/bastard/loser/liar/minx/brat/slut/squirt/sod/ bugger, etc.] [you]
2. Personalised negative assertions	– [you] [are] [so/such a] [shit/stink/thick/stupid/bitchy/bitch/hypocrite/ disappointment/gay/nuts/nuttier than a fruit cake/hopeless/pathetic/ fussy/terrible/fat/ugly/etc.] – [you] [can't do] [anything right/basic arithmetic/etc.] – [you] [disgust me/make me] [sick/etc.]
3. Personalised negative references	– [your] [stinking/little] [mouth/act/arse/body/corpse/hands/guts/trap/ breath/etc.]
4. Personalised third-person negative references (in the hearing of the target)	– [the] [daft] [bimbo] – [she] ['s] [nutzo]
Pointed criticisms/ complaints	– [that/this/it] [is/was] [absolutely/extraordinarily/unspeakably/etc.] [bad/rubbish/crap/ horrible/terrible/etc.]

Table 6.1 (continued)

Category	Formulae
Challenging or unpalatable questions and/or presuppositions	– why do you make my life impossible? – which lie are you telling me? – what's gone wrong now? – you want to argue with me or you want to go to jail?
Condescensions	– [that] ['s/is being] [babyish/childish/etc.]
Message enforcers	– listen here (preface) – you got [it/that]? (tag) – do you understand [me]? (tag)
Dismissals	– [go] [away] – [get] [lost/out] – [fuck/piss/shove] [off]
Silencers	– [shut] [it]/[your] [stinking/fucking/etc.] [mouth/face/trap/etc.] – shut [the fuck] up
Threats	– [I'll/I'm/we're] [gonna] [smash your face in/beat the shit out of you/box your ears/bust your fucking head off/straighten – you out/etc.] [if you don't] [X] – [X] [before I] [hit you/strangle you]
Negative expressives (e.g. curses, ill-wishes)	– [go] [to hell/hang yourself/fuck yourself] – [damn/fuck] [you]

It is important to note that, as Culpeper, Iganski and Sweiry (2017: 8) stress, such formulaic items by no means always have impolite meanings, nor are they impolite in all contexts. Yet they do have default interpretations as impolite, and in most usages *are likely* to demean the victim's identity (Walters 2013: 70).

The list in Table 6.1 includes forms common in English, but is in no way an exhaustive inventory of English impoliteness formulae. Furthermore, as Culpeper himself notes (2010: 3243), the linguistic repertoire of impoliteness formulae varies across languages (cf. Vasilaki 2014). In the present study, these categories will be tested on Lithuanian data to see how well this model can apply to a typologically different language. One major difference can be anticipated at the outset. In Lithuanian, a synthetic language, as opposed to English, an analytical language, a considerably larger diversity of morphological forms used to express impoliteness can be expected.

Having reviewed the most current pragmatic model of impoliteness, we can now compare which features impoliteness, in general, shares with hate speech in particular. These features are summarised in Table 6.2, which brings into comparison the defining characteristics of impoliteness with the six-part threshold test for criminally liable hate speech expressions, as set out in the Rabat Plan of Action (2012).

Table 6.2: Criteria for hate speech as a criminal offence and impoliteness.

Criteria for hate speech according to the Rabat Plan of Action (2012)	Impoliteness criteria according to Bousfield (2008: 124) and Culpeper (2010: 3233)
Context (social and political context prevalent at the time the speech was made and disseminated)	Context (expectations, desires and/or beliefs about social organisation)
Speaker (the speaker's position/status in the society and thus their power to incite crime against the target group)	Speaker and hearer (hearer perception/construction of the speaker's intent/hurtfulness of their words)
Intent (deliberateness to incite hatred against the target group)	Intent (deliberateness of the FTA)
Content and form (form, style, directness, nature of arguments deployed in the speech)	Content and form (conventional impoliteness strategies and formulae)
Extent of the speech act (the reach of the speech act, its public nature, its magnitude, and size of its audience)	Face-damaging effects (various factors can exacerbate how offensive an impolite behaviour is taken to be)
Likelihood, including imminence (probability that the speech would succeed in inciting actual action against the target group)	Emotional consequences (hearer face actually being, or not being, damaged)

Although impoliteness and hate speech are not precisely identical, the criteria in Table 6.2 show that they share characteristics. Impoliteness formulae could therefore indeed play an important role in helping classify hate speech for its automated identification online. The classes of impoliteness formulae in English data certainly need to be tested against data from other languages, in order to assess the general applicability of this model. However, it is anticipated that these categories will be useful in assessing the form and content of Lithuanian hostile online comments.

3 Data and methods

The data for this study was collected from Delfi,[4] a major Lithuanian news portal, which allows both registered and anonymous users to post comments. The entire dataset initially consisted of 734 comments (totalling 51,017 words) written in response to news reports on issues related to migrants. Since the focus of this research is on impoliteness forms in relation to hate speech, the dataset was refined by removing all comments without offensive content (totalling 135 comments) and by restricting it to abusive or discriminatory comments, which make up 599 comments (totalling 38,991 words).

The data was collected in 2018, a year marked by a sharp increase of forcibly displaced people and asylum-seekers, mainly from Syria, as a result of persecution, armed conflict, violence, or persistent human rights violations. According to the information in the United Nations High Commissioner for Refugees (UNHCR) review,[5] in 2018 the global population of forcibly displaced people increased by 2.3 million. By the end of the year, almost 70.8 million individuals were forcibly displaced worldwide. Although Lithuania hosted a low number of refugees, the current data shows that public debates on Lithuanian news portals were strongly affected with anti-refugee sentiment.

Abusive comments in the dataset express various degrees of hostility: offensive content (including derogatory language, obscenity, profanity and vulgarity), socially unacceptable content (attacking a target group, but not to a degree that is criminally liable), and potentially criminal hate speech.[6] In the present research, it was necessary to include all the various levels of hostility, in order to test whether the impoliteness framework helps disambiguate these levels, and to identify what qualifies as the gravest content and thus as potentially criminal.

As noted above, the present analysis applies Culpeper's (2010, 2011a, 2011b) impoliteness model. The categories distinguished in this model, and which will be examined in the present study, were summarised in Section 2. The analysis here combines qualitative and quantitative methods, but the qualitative approach prevails. First, the impoliteness formulae will be quantified, so as to establish their specific weight in the selected sample of anti-migrant comments. Second, these formulae will be examined for their specific function in the hostile argumentation in anti-migrant comments online.

---

4 https://www.delfi.lt/ (accessed 20 August 2022).
5 https://www.unhcr.org/globaltrends2018/ (accessed 5 September 2022).
6 For a more detailed explanation of these three levels of severity, see Nobata et al. (2016: 149) and Fišer, Ljubešić and Erjavec (2017).

Despite the apparent tidiness of the list, the classification and quantification of impoliteness formulae (overviewed in section 2) was not a straightforward task. Culpeper, Igansky and Sweiry provide a definition that establishes impolite items as conventionalised units, writing that "[a] conventionalised impoliteness formula is a regularly occurring bundle of language or a non-verbal sign in which context-specific offensive effects are encoded to a degree" (2017: 12). However, this very criterion of conventionality, regular usage, and encoded offensiveness is what made the status of impoliteness formulae sometimes difficult to determine. To assess conventionality, Culpeper (2011b) checked all of the items in his research in the two-billion-word Oxford English Corpus, which allowed him to ascertain that more than half of the instances of each type occurred in impoliteness contexts, which qualified such formulae as conventional. For the purposes of the present study, there was no way to check all the types against a general Lithuanian corpus, because the existing corpus is both out of date and is mainly restricted to more formal interactive contexts, where impoliteness is less often encountered. The datedness of the corpus, in particular, makes it of limited use for the present research, because hate speech is most often expressed in colloquial language, which tends to change relatively fast. Moreover, it would have been beyond the scope of the present research to examine all the instances of impoliteness formulae in general Lithuanian. In this study, the criterion of conventionality is therefore treated tentatively, relying mainly on native speaker intuition. Less conventional and unconventionally impolite forms have also been included, since hate speech tends to live in the ever-changing vernacular, and to resort to creativity and unconventional forms.

Another challenging problem was that of delineating the boundaries between impolite units, and deciding what should count as a distinct individual formula. For example, in cases of lists (especially when the list contains some synonymous words or words belonging to the same lexical set), it is uncertain how many formulae there occur. Consider example (1), where the speaker associates several criminal activities with migrants:

(1) *Mus sprogdina,traisko sunkvezimiais,saudo*[7]
 "They blow us up, crush us with trucks, shoot"

In cases like this, the entire list was considered a single formula, rather than treating each verb as an individual impoliteness unit.

---

[7] Note that the examples are presented without any corrections, leaving all the errors of orthography and grammar found in the original.

Yet another ambiguity arose when analysing the category of dismissals. In Culpeper's (2011a, 2011b) model, dismissals appear with the second person pronoun. In the data consulted for the present study, however, some dismissals were formulated with the third person pronoun (usually in the plural form), as in *tegul varo lauk* ('let them bug off'). Although such forms do not appear in Culpeper's (2011a, 2011b) catalogue of dismissals, in the current study, they are assigned to this category.

One of the most problematic distinctions was that between insults realised as third-person generalisations and those realised as pointed criticisms directed at a particular person. Because this issue requires more deliberation, it will be discussed below, in the analysis itself (sections 4.2 and 4.3).

In a number of comments, there is an interplay between more than one impoliteness category, and the same unit can sometimes be assigned to more than one category. In such instances, double labelling was applied, assigning such items to all the relevant categories.

As the research questions presented in the introduction (section 1) indicate, the analysis here will try to account for impoliteness formulae/strategies used in Lithuanian language online comments against migrants. Lithuanian impoliteness formulae will be generalised following Culpeper's (2010, 2011a, 2011b) model: different structural units will be presented in square brackets, and alternative items will be separated by slashes. The degree of conventionality will be partly grounded on the frequency of the formulae in the present data. By examining the alternative items in the formulae, the analysis will consider the degree of flexibility and creativity that these conventional structures allow for, under the assumption that automated identification of hateful content would be easier with more stable structures. Finally, the paper will provide some suggestions as to how the impoliteness framework can be improved and used more effectively for the purposes of manual and automated hate speech detection.

4 Analysis of impoliteness formulae in abusive online comments

The present section examines the use of different impoliteness categories. Figure 6.1 depicts the distribution of the raw frequency of each category, along with their proportions expressed in percentages. The most frequently occurring category is that of challenging or unpalatable questions and/or presuppositions, constituting 42% of all the impoliteness formulae in the data. The next most frequent category is that of insults (26% of all the formulae). All the other categories are at least two times less frequent (no more than 16%). Some of them, namely threats, enforcers, silencers and negative expressives, represent just a small minority of cases (no more than 1%).

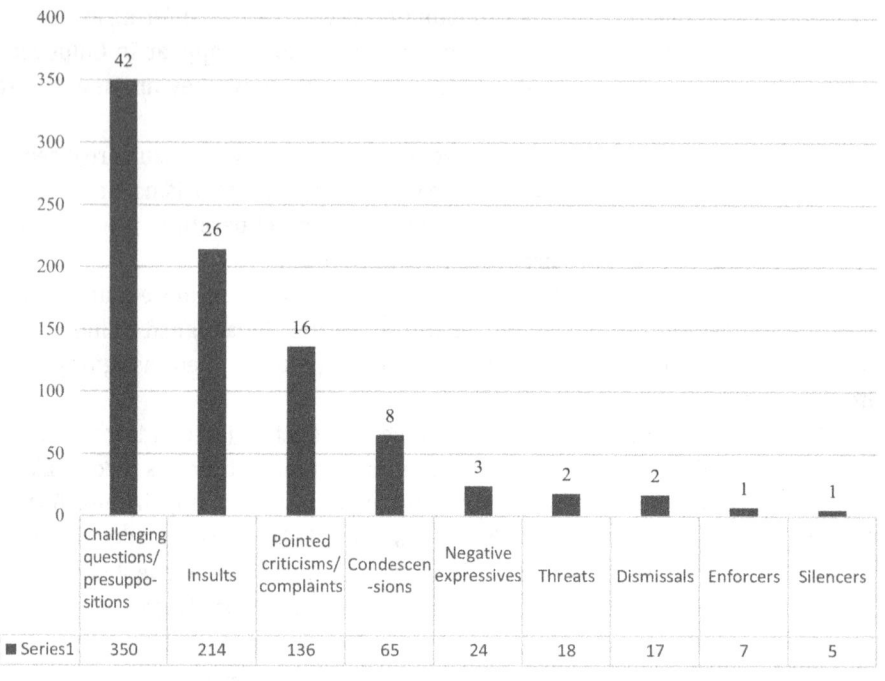

Figure 6.1: Frequency of impoliteness formulae in Lithuanian anti-migrant comments (raw frequency represented by the Y axis; percentages appear on top of each bar).

In Culpeper, Igansky and Sweiry's (2011: 15) study, in contrast, insults are the most frequent category (47%), followed by considerably less frequent threats (18%) and even less frequent negative expressives, incitement dismissals, commands, chants and other categories (3% to 8%). Challenging questions and presuppositions do not occur in their data, which suggests that they are not common in illegal hate speech but may tend to dominate in merely offensive content. Formulated in a non-declarative way, questions are used as an indirectness strategy and make discriminatory or attacking claims fit less well into specific impoliteness categories (cf. Chapter 7). The high incidence of challenging questions and presuppositions in the current data could also be influenced by the genre: the online comments are longer and more developed than the oral attacks examined by Culpeper, Igansky and Sweiry (2011).

The next section presents a more detailed discussion of the categories identified in the present research. They will be examined in their order of frequency, starting with challenging or unpalatable questions and presuppositions.

4.1 Challenging or unpalatable questions and presuppositions

As noted earlier, the category of challenging or unpalatable questions is the most frequent type of impoliteness, representing 42% of all impoliteness items in our data. A number of the comments contain more than one question and, to achieve a more emphatic effect, these questions are sometimes followed by more than one question mark (typically three). The largest number of questions that appears in a single comment is eleven (there are two comments with this number of questions).

Not only can some comments contain more than one question, but an entire comment may be formulated as a series of questions, as in example (2), which consists of five questions in a row:

(2)   "\"*Uzsienietiskos isvaizdos*" :D *Kur musu zodzio laisve, kur Europos laisvos spaudos vertybes, jei bijo ivardinti, kad imigrantai, kuriu prisileido, uzpuldinejo moteris? Kodel mes bijom kalbet savo salyse? Kas per cenzura? Gaunasi, kad gyvenam policinese, diktaturinese valstybese, nors garsiai kalbama apie demokratija, zodzio laisve? Kaip Europa iki to nusisrito?*
["\"of foreign appearance\" :D Where is our freedom of speech, where are the European values of free press if they are afraid to spell out that immigrants, who've been let in, were attacking women? Why are we afraid to speak up in our countries? What kind of censorship is that? It turns out that we live in police, dictatorial countries though we speak loudly about democracy, freedom of speech? How could Europe end up like this?]

When a comment is framed as a sequence of questions with no affirmative sentences, this reduces the degree of speakers' commitment to their underlying assumptions, and signals that these are not assertions but considerations.

Determining precisely which questions were challenging or unpalatable, and which were not, presented some difficulty in the analysis. To resolve this issue, a question was deemed challenging or unpalatable if, first of all, it aimed to criticise some social reality or development, and, secondly, if when reformulated as a statement, such a question would become a direct accusation or complaint about migrants. Example (3) is such a question:

(3) *Kodėl mes visi imigrantai sunkiai dirbdami turime jiems mokėti pašalpas? Tegu visi eina ir dirba, darbų ten netrūksta jei tik nori dirbti.*
[Why do all of us immigrants have to pay them benefits by working hard? Let everyone go and work, there is no shortage of jobs if only you want to work.]

Example (3) appeared in a discussion of Lithuanian migrants living outside Lithuania, where non-Muslim migrants are presented as "us", and Muslim migrants are stigmatised as "them". By using the question about state benefits, the commenter resorts to the common narrative of migrants as a social burden, but does it in a less direct way than if this interrogative were formulated as a declarative.

When questions contain negatively connoted references to people, the speaker's aim is to ridicule the target. Example (4) below includes a derogatory reference to women (using the Russian slang word *starucha* as an insult):

(4) *Savų sukriošusių staruchų pristigo?*
[Don't they have enough decrepit old women of their own kind?]

Formulated with a third person reference and as an interrogative, such comments appear as mitigated insults.

Challenging questions are often asked as rhetorical questions. In some instances, it is made explicit that they do not require an answer, since they are followed immediately by the speaker's own response, as in example (5).

(5) *kas nebijo išprotėjusios beždžionės su granata rankoj? Visi bijo. Tokia baimė—visiškai natūrali.*
[who isn't afraid of a mad monkey with a grenade in his hand? Everyone is afraid. Such fear is completely natural.]

As Shum and Lee (2013) observe, raising rhetorical questions is a common positive impoliteness output strategy (aiming to damage the addressee's positive face wants) in contexts of disagreement in online forums. In a similar vein, Kienpointner (2008: 259) observes that, in arguments between traffic wardens and owners of illegally parked cars, speakers were able to use a series of rhetorical questions to make accusatory implications against their interlocutors, a strategy which can attack both their positive and negative face. Vasilaki (2014: 101–104) relates rhetorical questions with insults, and observes that, in her data from Greek speakers, insults are often formulated as questions.

When someone asking a rhetorical question immediately follows it up with their own response to the question, they can express patronising attitudes, as in example (6).

(6) *Ka jus zinote apie tai kas darosi Amerikoje??? Nuli.*
 [What do you know about what is going on in America??? Nothing.]

Questions may serve as indirect commands or suggestions, as illustrated in example (7).

(7) *Gal galima ruskius i marsa isvezti? jei ne—tai reiks su jais gyvent salia, mes to noresim ar nenoresim.*
 [Maybe Russians can be sent to Mars? if not, we'll have to live with them side by side whether we want it or not.]

Questions create dialogicity. As such, questions are used to convey indignation at some debated issues, as in example (8).

(8) *islamistu motuses isakymu visi imigrantai nuo kovo vidurio, tie, kurie megins patekti i vokietija, bus issisti atgal i Graikija. Graikija??? O kodel ne namo??? Merkel bijo pralaimeti rinkimus del nezabotos imigracijos, taigi daro siaubingus sprendimus. Merkel i kaliuze uz zmoniu zudymus*
 [according to the order of the Islamist mother, all immigrants from mid-March, those who try to enter Germany, will be sent back to Greece. Greece??? And why not home??? Merkel is afraid of losing the election due to unbridled immigration, so she makes terrible decisions. Merkel to the nick for killing people]

The argumentative style in example (8) is based on the following sequence: (1) raising an issue, (2) asking challenging questions regarding it and (3) responding to their own rhetorical questions in a hostile manner.

Generally, just as in the examples discussed above, comments involving challenging questions express hostility, discriminatory attitudes and humiliating claims about migrants, but the question form itself serves as redressive action to soften the face-threatening action (FTA). This is presumably done not so much to counteract possible face damage to the hearer, but to diminish the speaker's commitment to the attitudes expressed, thus avoiding moral or legal responsibility.

4.2 Insults

Insults can be defined as "derogatory nominations" and include name-calling through the use of negatively connoted epithets, taboo words and swearing (Culpeper 2011a; Vasilaki 2014). In the present data they constitute the second most frequent category, but not all types of insults are equally distributed. As shown in Table 6.3, one type of insults clearly outweighs the other types, that of third-person negative references, which make up 80% of all the insults in the dataset. The categories of personalised negative assertions (12%) and personalised negative vocatives (7%) are considerably lower in frequency.

Table 6.3: Types of insults in hostile online comments targeting migrants.

Type of insult	Example	Frequency	%
Third-person negative references	*Pazekit kas dedasi eu kai isileido **pirmykscius zmones**. Kiek esu dirbes su **negrais** ir **babajais visi jie niekam tike**.* "Look at what's going on in the EU after they let in **primeval people. All the negros and babaji** I've worked with **are good for nothing**."	173	81
Personalised negative assertions	*Norėčiau, kad tave parurintu. **Tu**, be abejo, **rurinies su babajais**.* "I wish you got banged. **You**, of course, **bang babaji**."	26	12
Personalised negative vocatives	***babajai**, welcom i tranzitine sali, labai dziaugiames jus cia matydami. Dar daugiau dziaugsimes, kai patepsit slides pas merkeliene.* "**babaji**, welcome to the transit country, we're very glad to see you here. We'll be even gladder when you take a hike to Merkel."	15	7
Personalised negative references	–	0	0
Total:		214	100

Interestingly, as shown in Table 6.3, there are no instances of personalised negative references, possibly due to the limited size of the dataset, sociocultural factors or language-specific reasons. Therefore, to learn more about the usage of this category of impoliteness strategy, further research based on a larger dataset is needed. In the following section, I turn to the more frequent categories of insults, and start with insults realised as third-person negative references as the dominant category.

4.2.1 Insults realised as third-person negative references

Distinguishing this category of insult from pointed criticism posed some challenges, a matter to which I will return at the end of section 4.2 and in section 4.3. Third-person negative reference insults were ultimately defined according to two criteria. Firstly, they should include a negatively loaded epithet to refer to the target, such as the items in bold in examples (9) and (10). Secondly, it should be likely or at least technically possible for the insult to reach the target. In this, the definition follows Culpeper, Igansky and Sweiry (2017: 17), who define insults as occurring in the presence of the hearer, whereas pointed criticism targets an absent third party.[8] In practice, comments such as those in (9) and (10), could plausibly be accessed by migrants or Muslim community members, thanks to the fact that they are likely to remain accessible there in perpetuity unless they are removed a moderator for their strongly abusive content. Nonetheless, the second criterion is admittedly speculative and debatable.

Insults realised as third-person negative references in this dataset are formulated mainly to promote the myth of "migrants as a threat" (cf. Chapter 8, this volume). Structurally, they include diverse patterns (which makes it difficult to generalise them as well-defined formulae). However, one observably dominant trend is that they include highly degrading vulgarisms and criminalise or dehumanise migrants (literally and metaphorically) by making downgrading references to them in the third person, as can be seen in the insults highlighted in bold in examples (9) and (10).

(9)   **musulmonai teroristai** <u>susaudo zmones</u> vidury miesto paryziuj, vokietijoj <u>mergas prievartauja</u>
      [**Muslim terrorists** <u>shoot people dead</u> in the middle of paris, <u>rape girls</u> in Germany]

(10) **Tie babajai** <u>tingi dirbti</u>, visur europoje <u>slankioja dyki</u>, nieko neveikia.
      [**Those babaji** <u>are too lazy to work</u>, all over Europe they <u>hang around doing nothing</u>.]

Alongside the insulting nominations, Muslims are criminalised, with accusations of murder and rape, as seen in example (9), and defamed as lazy, as in example (10).

---

8 In Culpeper, Igansky & Sweiry's (2017: 17) model, the hearer's presence is a crucial criterion in distinguishing two impoliteness strategies: pointed criticism and insults formulated as third-person references. This criterion was important and relevant in their framework because their taxonomy was developed on the basis of spoken data and applied to examine oral communication, where the hearer's presence is important (differently from asynchronous digital communication).

Viewed from a discourse analytic perspective (cf. Leeuwen 2007), such heavily intensified, overgeneralising and misinforming claims serve as a legitimation strategy, justifying hate and hostility. These pseudo-factual claims are formulated as factual and as universal by using universal pronouns (*visi* 'all') and adverbs (*visur* 'all over') and present tense verb forms, underlined in examples (9) and (10), without any hedging or mitigation.

4.2.2 Personalised negative assertions

Personalised negative assertions address a targeted individual in the second person (*tu* 'you') and/or the verb in the second person, as in examples (11)–(13). In reading the following examples, note that in Lithuanian, the use of the personal pronoun in such structures is optional, since verbs are obligatorily marked for person by using inflections—as in *esi* 'you are' in example (12), where the ending *-i* marks the second person singular form. Thus, the subject noun or pronoun may be omitted in Lithuanian as redundant.

(11) *Norėčiau, kad tave parurintu.* ***Tu****, be abejo,* ***rurinies su babajais****.*
[I wish you got banged. **You**, of course, **bang babaji**.]

(12) *Esi tikriausia ožio ir beždžionės sueities vaisius.*
[**You are** perhaps **a foetus of a sexual intercourse between a goat and a monkey.**]

(13) *... AČIŪ,* ***tau****, Angele, kad iš EU...... tu padarei—PROCHODNOY DVOR...... kur ne tik visokio plauko \"pabėgėliai\", bet...... ir putinfiureris gali veikti, KĄ TIK NORI....... DANKE NICHT schone"*
[... THANK **you**, Angele, for turning the EU ...... into—THE TRANSITIONAL YARD ...... where not only the \"refugees"\ of all sorts, but ...... also putinführer can do WHATEVER THEY WISH .... DANKE NICHT schone]

Example (11) also falls into the category of ill-wishes (see section 4.4), and example (13), which includes a conventional politeness formula for thanking but does not meet sincerity conditions,[9] is an instance of mock politeness. These are good examples of how multiple impoliteness strategies can be used in the same utterance.

---

9 When sincerity conditions are met, the utterance accurately represents the speaker's intention.

Importantly, in examples (11)–(13), as in many other comments in the dataset, commenters do not target migrants, but instead direct their hostility against those supporting them. The targeted supporters can be other commenters, as in examples (11) and (12), or politicians seen as being responsible for refugee-friendly policies, as in example (13), where the former Chancellor of Germany Angela Merkel is addressed with the Lithuanianised name *Angele*, which is used pejoratively.

Note that in example (13) there are two elements in languages other than Lithuanian: the Russian saying *PROCHODNOY DVOR* ('TRANSITIONAL YARD') and the (deliberately incorrect) German thanking formula *DANKE NICHT schone* ('THANK YOU NOT very much'. The use of foreign linguistic elements of different size, such as affixes, words, and phrases, could be a research topic of its own, as each seems to contribute to meaning in different ways. For instance, they can be treated as evidence of the speaker's creativity (especially in the case of neologisms), and can emphasise emotionally loaded content.

In cases such as those presented in examples (11)–(13), impoliteness formulae are used to attack the target's face and hence may have face-damaging effects. In so doing, they may create a toxic discursive environment, silence differently minded participants in the discussion and may lead to conflictive communication by demeaning the target's identity. However, according to Lithuanian legislation, such comments are unlikely to qualify as legally punishable hate speech, because the severity of such attacks, and the probability that such speech would succeed in inciting actual action against the target group, are relatively moderate. Such utterances, characterised by impoliteness formulae, are socially unacceptable verbal behaviour, but so far they do not seem to be positively definable as criminal hate speech.

4.2.3 Personalised negative vocatives

Lithuanian impoliteness formulae in this category include structures with *tu/jūs* ('you' with the T-V distinction) and an address form; they follow such patterns as the examples below:
- asilo [tu] galva ('you donkey's head')
- [tu] nevala ('you scumbag')
- [jūs] pabėgėliai ('you refugees')
- [ei] babajai ('hey babaji')

The derogatory term *babaji*, used here to refer to Muslims, is highly conventionalised and refers to this specific target group in an unambiguously derogatory way. The vocatives *asilo tu galva* ('you donkey's head') and *nevala* ('scumbag') are conventionally impolite, but can be used to address a wider range of target groups. The negativity of *pabėgėliai* ('refugees') emerges only in context, but in the present dataset the negative context prevails.

Thus, personalised negative vocatives include nouns, which can be realised through the following categories:
- Derogatory noun in the Vocative case: *liurbi/arogantiški trumpapickiai/lochai/ nevykėle/gudruti* ('sucker'/'arrogant nincompoops'/'dorks'/'loser'/'genius', the latter being ironic and intended to mean its opposite)
- Derogatory noun + postmodifier: *žmogžudy, beždžionės pavidalu atrodantis* ('murderer looking like a monkey')
- Patronising vocatives (realised by diminutives): *vaikeli/kvailele* ('kiddie'/'little fool')

As shown in the examples above, many negative vocatives are synonyms of "a fool" (cf. Ruzaitė [2021: 135] for similar findings in topically diverse offensive comments). Generally speaking, personalised negative assertions and negative vocatives are less frequent in hostile comments online, but they appear to be more face-threatening than third-person references. The high frequency of insults is also reported in Culpeper (2011a) and Culpeper, Igansky and Sweiry (2017: 16–18) where they observe that, in their data, insults heavily outweigh all other categories of impoliteness. In contrast with the findings in the present study, the two types of insults that dominate in Culpeper, Igansky and Sweiry (2017: 16–18) are personalised negative vocatives and personalised negative assertions. However, in Kleinke and Bös' research (2015), which examined intergroup rudeness in English and German online discussion fora, the most frequently occurring category is pointed criticism.

The distinction between insults realised as personalised third-person negative references and pointed criticisms/complaints is imprecise. As mentioned earlier, pointed criticism, according to Culpeper, Igansky and Sweiry (2017: 17), targets a third party, whereas insults target the addressee's face. The target is more explicit in personalised insults with the pronoun *you*, but when third person references are used as insults, there is inevitable ambiguity in possible interpretations.

Regarding this distinction, Culpeper, Igansky and Sweiry (2017: 16–18) admit that there is overlap between the two categories, and they stress that context becomes crucially important in classification. If the group targeted in an utterance is present when a third-person generalisation is made, then this is an instance of an insult. But if the claim is made when the target is not present in the context of situation, it is an instance of pointed criticism. This distinction was formulated with reference to face-to-face interaction, and is not easily applicable in asynchronous digital interchanges. Since digital interactions are marked by the feature of longevity[10] (Graham & Hardaker 2017: 789–790), in the present study it is impossible to know what audience the comment reached nor whether the commenter intended to reach the target group. In many cases, it is simply not possible to disambiguate the two categories, which makes it difficult to compare the quantitative results of different studies, and challenging to quantify such fuzzy categories even within the scope of a single study.

4.3 Pointed criticisms/complaints

To continue the discussion of the distinction between third-person generalisations as insults and pointed criticisms, I now turn to an overview of the latter category. In the present study, generalisations about individuals or groups of people are classified as pointed criticism if they are not likely to reach their target, as in example (14), where references are made to former Chancellor of Germany Angela Merkel.

(14) *O ka Vasingtonui kalbet su nukvakusia boba,kuri per savo durnuma uztvinde Vokietija babajais ir baqia suzlugdyt Vokietija*
[And what should Washington talk about with the crazy woman who through her stupidity flooded Germany with babaji and will soon finish ruining Germany]

When high-ranking politicians are referred to in online comments, it is unlikely that the commenter actually intends to reach those people, nor expects that the targeted persons will actually read such comments. The main goal is to criticise and complain about the targeted individual, target group, various events, phenomena, or about the general status quo. In addition, in cases where it was impossible to assess the likelihood of the comment reaching its target, and if they were realised mainly as degrading labels or epithets, the examples were categorised as insults. By contrast, pointed criticism includes comments realised through predicational strategies, mainly through references to some actions or qualities of the target, as in examples (9) and (10) discussed in section 4.2. In these examples, the underlined items (verb phrases) are classified as pointed criticisms and complaints, and the items in bold (nominal phrases) are classified as insults.

---

10 If items posted online are not reported as abusive content and are not removed by a moderator, they are usually accessible for a long time after their initial posting.

4.4 Some other (minor) impoliteness strategies

The least frequent impoliteness categories encountered in this research include condescensions, patronising, curses/ill-wishes, dismissals, threats, enforcers and silencers. Since these are rarely used, they will be discussed here only briefly, by highlighting the main aspects of their usage.

4.4.1 Condescensions

Condescensions and patronising include belittling, ridiculing and demeaning behaviours "[p]roducing or perceiving a display of power that infringes an understood power hierarchy" (Culpeper 2010: 3240). In Lithuanian anti-migrant comments, patronising attitudes and disdain are realised through morphological and lexical means of expression.

Morphologically, one important form used to realise pejorative attitudes is that of diminutives. In Lithuanian, there is a large diversity of diminutive suffixes, which are used extensively in a variety of social contexts, usually less formal ones. Although diminutives typically express endearment (e.g. in child-directed speech), in hostile online comments they serve the opposite function, expressing negative attitudes. See example (15).

(15) *Vaik**eli**, bet kokiame universitete tau net labai neišsilavinęs dėstytojas paaiškins, kad nuorodos į Wikipedija nepriimamos*
[Kidd**ie**, at any university even a teacher who is not very educated will explain to you that links to Wikipedia are not accepted]

In fact, not only is the diminutive form central here, but lexical choice is also important. As a vocative, "Kiddie" functions as a power display, through which the addressee is downgraded. As for the use of suffixes to express attitudinal meanings, the findings here are in line with Vasilaki (2014: 101), who indicates that diminutives are used in Greek insults to belittle the target and thereby damage the addressee's positive and negative face.

Apart from some specific nouns, such as the diminutive address forms, other lexical elements can be used to make patronising comments online. In the present data, contempt toward the hearer (migrants, their supporters or any participant showing disagreement) is expressed through allusions to the circus (e.g. *Cirkai!!* 'Circus!!') or by using the imperative form *nejuokink* ('don't make me laugh') and different variations of this verb, such as *Nejuokink(it)* ('don't make me laugh'), *Nebejuokink(it)* ('stop making me laugh') and *Varykit namo nejuokinę (zmoniu)* ('Go home without making [people] laugh'). In such instances, othering is achieved by belittling the Other, treating them as an object of ridicule.

Finally, patronising is often expressed in face-threatening actions (FTAs) by explicitly disparaging the addressee's intellectual capacities, e.g. *nieko jus neikertat* ('you don't get anything'). Kleinke and Bös (2015: 63), similarly, point to the large number of negative comments on the intellectual capacities of other online posters in a German discussion forum, in contrast with the paucity of such face-attacks in English forum discussions.

4.4.2 Dismissals

The typical formulae for dismissals are as follows: [Tegu ('Let')] [*verb*] and [Tai ('So')] + [*imperative*]; for instance, take example (16) below.

(16) **tai eik draugauk su babajais**, *pažiuresim kaip seksis asilo tu galva*
[**so go and be friends with babaji**, we'll see how you'll get along you donkey's head]

As shown in example (16), in dismissals the target is addressed directly. In addition, the utterance is aimed at banishing the addressee from the speaker's social space through the use of swearing.

4.4.3 Negative expressives

Negative expressives mainly include curses and ill-wishes. They are similar to dismissals in their form, because they can start with the particle *tegul/tegu*, as in example (17). Another form used to express dismissals is the verb *norėti* in the conditional form *norėčiau* ('I wish/would like'), as in example (18).

(17) *Ogi* **tegu sauna sau i p.autus**. *Kojas abi persisove. R.roj sulinde jau imigrantaiXa xa xa*
[**Let them shoot in their own balls**. They've already shot themselves in the foot. Immigrants are already stuck in the ass Ha ha ha]

(18) *Norėčiau, kad tave parurintu. Tu, be abejo, rurinies su babajais.*
[**I wish you got banged**. You, of course, bang babaji.]

As examples (17)–(18) suggest, dismissals tend to include vulgarities. Shum and Lee (2013) classify these as a common impoliteness form.

4.4.4 Threats

Threats, according to Culpeper (2011a: 136), can be "conditional"[11] (providing a reason for acting against the target group) or "non-conditional" (providing no reason for acting against the target group). However, based on those criteria, Culpeper's (2011a) data only includes examples of conditional threats (cf. Limberg 2009 for similar results). In contrast, in later research on religiously aggravated crimes, Culpeper, Igansky and Sweiry (2017: 18–19) identified no instances of conditional threats whatsoever. In the present study of Lithuanian anti-migrant hostility, 18 threats were found, only two of which are non-conditional. Both types of threats, conditional and non-conditional, can be performed explicitly or implicitly.

Threats are used for addressing and threatening migrants or opponents in the discussion. They are used to frighten migrants into believing that they can be victims of harmful acts at some indefinite time in the future. Threats can be formulated as a call for violent action against the target group, as in example (19). The call for violent action against the target may meet some of the legal criteria for sanctioned hate speech in the European Union.

(19) ***Ginkluokimes**, nes babajai (emigrantai) ar alkani bei nusivile viskuo buduliai ateityje tik suaktyves*
 [Let's arm ourselves because babaji (emigrants) or hungry and disappointed brats will only get more active in the future.]

Example (19) illustrates a conditional threat where the reason for acting against the target group is provided in the subordinate clause.

Some threats against migrants are less explicit, as in example (20), where a generalising claim is made by using a third person reference to migrants without addressing them directly. However, the implied meaning clearly points to a call for physical violence. Here we can also see how a threat is used alongside a negative expressive; the latter is expressed by making a reference to minefields.

---

[11] In Culpeper's works (2011a), the term "conditional" is not used to refer to the syntactic structure of an utterance and does not mean that the threat is necessarily formulated as a conditional clause. In his model, this term refers to whether a message provides a reason or justification for hostility against a target group or not.

(20) *Kova su nelegalais yra pigi ir paprasta tai minu laukai, o cia matai duokit babkiu* **as padarysiu posol jautriai n......**
[Fighting illegal immigrants is cheap and easy these are minefields, and here you see they ask for money **I'll make it so that you are f***ed up gently**]

Example (20) contains an implicit reference to the condition or reason for the commenter's hatred of migrants, i.e. the reference to benefits. Although the comment is offensive, it could not be legally sanctioned unless it could be demonstrated that the comment is likely to succeed in inciting actual violent action against the target group. Implicitness thus reduces the severity of the comment, which otherwise clearly discriminates against the target group and frames it as an object of potential violence.

A non-conditional threat is provided in example (21). Again, the comment is not sufficiently explicit to qualify as legally punishable, but it does voice support for violence.

(21) *nieko komuniagos* **jums nebeilgai liko**
[no worries communists **you don't have long left**]

As mentioned above, impoliteness forms, including threats, are often directed against those supporting migrants, e.g. politicians, journalists or other commenters, as illustrated in example (22).

(22) *Merkel i kaliuze uz zmoniu zudymus*
[Merkel to prison for murdering people]

Here a conditional threat is used; the reason for the threat is provided in the prepositional phrase *uz zmoniu zudymus* ('for murdering people').
In some comments, threats are formulated as imperatives or commands with the modal verb *reikia* ('need'/'it is necessary'); these are listed below.
- *Reikia* vienytis ir daryti revoliucija ('We need to unite and make a revolution')
- *Reikia* tokias uzdaryt ('They need to be locked up')
- *reikia* sustabydi imigraciją ('immigration needs to be stopped')
- *reikia* juos vyti ir kuo greiciau ('they need to be driven out as soon as possible')
- *Reikia* atsikratyti parazitų ('we need to get rid of the parasites')
- nuo jų *reikia* Gintis ('we need to defend ourselves from them')

- Mldc, taip ir *reikia* elgtis su tais babajais-tegul varo is LT lauk ('Well done, that's how those babaji need to be treated-let them disappear from Lithuania')
- kazka daryti *reikia* su nelegalais ('something needs to be done with aliens')
- Kaip su tokiais ir *reikia* pasielgti—supakuoti ir atgal. ('That's exactly what needs to be done with them—pack them and send them back').

Whether expressed implicitly or explicitly, commands as directive speech acts can be considered a type of threat when they are formulated as warnings, or when they encourage addressees to act against a target group and aim to display power (cf. Culpeper et al. 2017: 19). In Lithuanian online comments, commands are not personalised because they are formulated in the third person rather than in the second person, as would be expected from Culpeper's (2010: 3242–3243) model. Still, in the present study, although they are categorised as threats, they are understood to be less direct than those formulated with the second-person forms. For instance, if the threat "they need to be driven out as soon as possible" were formulated not with the third-person reference but with the second person ("you need to drive them out as soon as possible"), the latter formulation would constitute a direct call for action.

Because of the relatively low number of threats found in this research, it is impossible to generalise that hateful commenters in Lithuanian use only threats expressed through the modal of necessity and through imperative forms. To draw any firm conclusions, it would be necessary to repeat the analysis with a larger sample and compare the results. Such a comparative study might find that the syntactic flexibility of Lithuanian conditions a high degree of flexibility in such formulations, making it difficult to reduce them to a clear set of formulae.

4.4.5 Enforcers and silencers

Enforcers and silencers are minor categories in the data. Each constitutes only 1% of all the impoliteness forms used, but they are equally good indicators of the language used by hate-advocating speakers.

Enforcers are used to enhance the pragmatic force of an utterance and increase its intensity. Structurally, they can be formulated as alerts, prefaces and tags. When used as alerts, they express a warning against some danger allegedly posed by migrants, e.g. *Atsibuskite!* ('Wake up!'). The alert then includes a verb in the imperative form, takes the plural ending for the second person to address a broad audience, and is followed by an exclamation mark. Such an utterance functions as a directive speech act and urges the hearers to act.

Enforcers can also be used as prefaces, as in *[Nu] [rimtai]* ('But seriously'). The item *nu* can perform versatile pragmatic functions in Lithuanian, and can be used as an interjection or a particle. Importantly, *nu* is often used as a discourse marker in informal communication to signal the beginning of an utterance. Followed by the adverb *rimtai* ('seriously'), it draws attention to the upcoming message and thus strengthens its impact.

Another way to express an enforcer is through a tag, e.g. *nesupranti* ('don't you get it'). Such forms intensify the preceding argumentation, and at the same time express the speaker's patronising attitude.

Silencers are explicit directive speech acts aiming to discourage the addressee from participating in the discussion. The silencer formulae found in the present data are listed below.
- *[kam jau kam, o] [prancuzijai su savo babajais] [geriau patylėti]* ('out of all countries, France with all their babaji should speak last')
- *Neskieskit [apie uzsienius]* ('don't talk rubbish about living abroad')

One of the alarming effects of hate speech is that its prevalence can silence the differently minded. In this respect, silencers are important, as they represent explicit attempts to suppress unwanted voices.

In summary, the categories examined in this section (section 4.4.) are small in number, but some of them, especially threats and ill-wishes, can be useful markers of hate speech. To discuss the importance of all the categories identified, I now turn to some concluding generalisations in relation to the research questions posed in this study.

5 Conclusion

This chapter has examined the categories of impoliteness found in Lithuanian online comments against migrants. The purpose was to determine whether or not impoliteness strategies could be adopted as potential classifiers of hate speech in Lithuanian, possibly for eventual use in the automated detection of hate speech. The study addressed three main research questions: (1) What impoliteness formulae/strategies are used in anti-migrant Lithuanian comments?, (2) How much flexibility and creativity do these conventional structures permit?, and (3) What remains unaccounted for, if only conventional impoliteness strategies are considered?

Regarding (1), this study has determined that impoliteness strategies in Lithuanian anti-migrant comments online vary considerably in their frequency. First, challenging or unpalatable questions and presuppositions are the most frequent category (42%). They are used as an indirect way to present some accusatory and discriminatory claims about the target, and thus can be useful when exploring discriminatory discourses in general. However, in most cases, the use of such questions or presuppositions is insufficient for a comment to meet the legally sanctionable criterion of criminal hate speech in content and form. Indirectly formulated claims about a target group are less likely to qualify as legally punishable hate speech (for an exploration of covert hate speech, see also Chapter 7).

Second, insults, which include personalised negative vocatives, personalised negative assertions and third-person negative references, make up 26% of all impoliteness categories. Insults realised as third-person negative references dominate (80%). They mainly promote the myth of migrants as a threat (cf. Chapter 8), include highly degrading vulgarisms, or criminalise and dehumanise migrants (literally and metaphorically). Negative vocatives and personalised negative assertions are less frequent but are more face-threatening than third-person references and, therefore, are more indicative of hate speech. Insults, in the same way as challenging questions, are arguably the easiest forms to pinpoint either manually or automatically.

A number of other considerably less frequent impoliteness strategies observed in the comments include: condescensions, dismissals, threats, curses/ill-wishes, silencers and message enforcers. However, they are not sufficiently pervasive in the current data to make broader generalisations about their prevalence, function or meaning. What is noteworthy, though, is that their infrequent use does not mean that they are insignificant when assessing whether a comment qualifies as sanctionable hate speech. On the contrary, it can be argued that, for instance, threats and ill-wishes can be more clearly indicative of hate speech than other more frequent, but also more ambiguous, categories.

Findings of the present research suggest that the impoliteness model used here, based mainly on the work of Culpeper (2010, 2011a, 2011b), lacks straightforward categories that could help to reliably differentiate between legally sanctionable hate speech and merely offensive content. On the other hand, there are no specific language forms that unambiguously serve as classifiers of hate speech.

Even though impoliteness categories mainly indicate the presence of biased attitudes and acts of bias, they can also serve as useful identifiers of offensive content and high-risk texts. Some systems used for automated hate speech detection are based on the principle of measuring the overall degree of toxicity in texts, thereby pinpointing high-risk texts which need to be further processed manually. An example of such a model is Perspective API (Salminen et al. 2020), which is a machine learning model developed to identify abusive comments online, on the basis of several attributes (e.g. Severe Toxicity, Insult, Threat and

Profanity). The attributes used in this model seem to overlap to a large extent with the categories used in impoliteness research.

As for RQ2, which aimed to examine the internal structure of impoliteness formulae and determine what is flexible or stable in them, this study found that there is little stability in impolite structures in Lithuanian. Although Culpeper's (2011a) impoliteness model focuses on conventionally impolite forms, findings from the present study tentatively point to more flexibility in the impolite forms used. The flexibility in the data may be partly explained by the creativity typical of hate speech. In addition, structural flexibility is at least partly shaped by the syntactic and morphological variation characteristic of synthetic languages such as Lithuanian. As a result, it is difficult to reduce impoliteness expressions in Lithuanian to clear formulae. Such findings may suggest limited applicability of impoliteness categories for automated identification of hateful content. On the other hand, the lack of clear formulae might also be due to the limited amount of data accessed in the present study. To reach more definite conclusions, it would be necessary to work from a larger dataset.

As regards RQ3, which addressed the issue of what remains unaccounted for if only conventional impoliteness strategies are considered in analysing hostile online comments, the results suggest that impoliteness formulae form an intricate texture in Internet comments, and are part of complex constellations of narratives (or *topoi*). The argumentation strategies in them are based on quasi-rational arguments, pseudo-factual information, legitimation strategies and discriminatory strategies (e.g. scare tactics, polarisation, othering and dehumanising). In addition, hate speech often appears in disguise (e.g. through intricate argumentation, irony and humour), and thus requires close reading, manual analysis and a holistic approach. All these aspects can be addressed by combining impoliteness categories with Critical Discourse Analysis, which can offer suitable frameworks for examining argumentation strategies more exhaustively.

What remains unresolved is the fuzziness of third-person generalisations, which can be classified as insults or pointed criticism. The present study tried to adhere to this distinction in the empirical analysis by drawing a line between strategies of nomination and predication, and by considering the question of the hearer's involvement. However, the results suggest that the latter criterion, and perhaps this categorisation in general, while relevant in face-to-face interactions is hardly applicable in digital interactions. Hence, these two categories should either be merged into one when examining this mode of communication, or finer criteria, more sensitive to the specifics of online communication, should be developed to distinguish them.

What remains unexplored in this study are intensification of hate speech, addressed systematically in Culpeper, Iganski and Sweiry (2017), and the use of irony, which in some research on impoliteness has been established as an important impoliteness strategy. For instance, Tselika's (2015) paper on irony determines that it is an important impoliteness tool, one which has a function of aggravating face-damaging effects. Similarly, Shum and Lee (2013: 67) explore ironic statements as a strategy of disagreement in Hong Kong Internet discussion fora. Thus, the use of irony in cases of hate speech and intensification of hate speech (only touched upon in this study) could be topics of future research.

Finally, the scope of all the generalisations in this chapter is limited to the data collected for this study alone. For this reason, the findings here cannot be extrapolated to generalise about the language of online commenters overall. The number and type of categories identified may have also been impacted by the group targeted in the comments, i.e. migrants, and it may be expected that attacks against other groups would involve different impoliteness forms and content. As a last word, in order to make any broader generalisations about Lithuanian hate speech online, more extensive and varied data are needed.

References

Bolinger, Renée Jorgensen. 2015. The pragmatics of slurs. *Noûs* 51(3). 439–462.
Bousfield, Derek. 2008. *Impoliteness in interaction*. Amsterdam: John Benjamins.
Brown, Penelope & Levinson, Stephen C. 1987. *Politeness: Some universals in language usage*. Cambridge: Cambridge University Press.
Carney, Terrence. 2014. Being (im)polite: A forensic linguistic approach to interpreting a hate speech case. *Language Matters* 45(3). 325–341.
Craig, Robert, Karen Tracy & Spisak Frances. 1986. The discourse of requests: Assessment of a politeness approach. *Human Communication Research* 12(4). 437–468.
Croom, Adam M. 2015. Slurs and stereotypes for Italian Americans: A context-sensitive account of derogation and appropriation. *Journal of Pragmatics* 81(1). 36–51.
Culpeper, Jonathan. 1996. Towards an anatomy of impoliteness. *Journal of Pragmatics* 25(3). 349–367.
Culpeper, Jonathan. 2005. Impoliteness and entertainment in the television quiz show The Weakest Link. *Journal of Politeness Research: Language, Behaviour, Culture* 1(1). 35–72.
Culpeper, Jonathan. 2009. The metalanguage of impoliteness: Explorations in the Oxford English Corpus. In Paul Baker (ed.), *Contemporary Corpus Linguistics*, 64–86 London: Continuum.
Culpeper, Jonathan. 2010. Conventionalised impoliteness formulae. *Journal of Pragmatics* 42(12). 3232–3245.
Culpeper, Jonathan. 2011a. *Impoliteness: Using language to cause offence*. Cambridge:Cambridge University Press.
Culpeper, Jonathan. 2011b. Politeness and impoliteness. In Karin Aijmer & Gisle Andersen (eds.), *Sociopragmatics*, 391–436. (Handbooks of Pragmatics 5). Berlin/Boston: De Gruyter Mouton.
Culpeper, Jonathan. 2013. Impoliteness. In Jan-Ola Östman & Jef Verschueren (eds.), *Handbook of pragmatics*, 1–18. Amsterdam: John Benjamins.

Culpeper, Jonathan, Derek Bousfield & Anne Wichmann. 2003. Impoliteness revisited: With special reference to dynamic and prosodic aspects. *Journal of Pragmatics* 35(10–11). 1545–1579.
Culpeper, Jonathan & Claire Hardaker. 2017. Impoliteness. In Jonathan Culpeper, Michael Haugh & Dániel Kádár (eds.), *The Palgrave handbook of linguistic (im)politeness*, 199–226. London: Palgrave.
Culpeper, Jonathan, Paul Iganski & Abe Sweiry. 2017. Linguistic impoliteness and religiously aggravated hate crime in England and Wales. *Journal of Language Aggression and Conflict* 5(1). 1–29.
Fišer, Darja, Nikola Ljubešić & Tomaž Erjavec. 2017. Legal framework, dataset and annotation schema for socially unacceptable online discourse practices in Slovene. *First workshop on abusive language online*, 46–51. Vancouver: Association for Computational Linguistics.
Goffman, Erving. 1967. *Interactional ritual: Essays on face-to-face behavior*. New York: Anchor Books.
Graham, Sage L. & Claire Hardaker. 2017. (Im)politeness in digital communication. In Jonathan Culpeper, Michael Haugh & Dániel Kádár (eds.), *The Palgrave handbook of linguistic (im)politeness*, 785–814. London: Palgrave.
Grainger, K. 2011. 'First order' and 'second order' politeness: Institutional and intercultural contexts. In Linguistic Politeness Research Group (ed.), *Discursive approaches to politeness*, 167–188. Berlin/Boston: De Gruyter Mouton.
Haugh, Michael. 2015. Impoliteness and taking offence in initial interactions. *Journal of Pragmatics* 86(1). 32–42.
Haugh, Michael & Valeria Sinkeviciute. 2019. Offence and conflict talk. In Matthew Evans, Lesley Jeffries & Jim O'Driscoll (eds.), *The Routledge handbook of language in conflict*, 195–213. London: Routledge.
Kienpointner, Manfred. 2008. Impoliteness and emotional arguments. *Journal of Politeness Research: Language, Behaviour, Culture* 4(2). 243–265.
Kleinke, Sonja & Birte Bös. 2015. Intergroup rudeness and the metapragmatics of its negotiation in online discussion fora. *Pragmatics* 25(1). 47–71.
Lakoff, Robin. 1977. What you can do with words: Politeness, pragmatics and performatives. In Andy Rogers, Bob Wall & John P. Murphy (eds.), *Proceedings of the Texas conference on performatives, presuppositions and implicatures*, 79–106. Arlington, VA: Center for Applied Linguistics.
Lakoff, Robin, 1989. The limits of politeness: Therapeutic and courtroom discourse. *Multilingua* 8 (2–3). 101–129.
Langlotz, Andreas & Miriam A. Locher. 2017. (Im)politeness and emotion. In Jonathan Culpeper, Michael Haugh & Dániel Kádár (eds.), *The Palgrave handbook of linguistic (im)politeness*, 287–322. London: Palgrave Macmillan.
Leech, Geoffrey N. 1983. *Principles of pragmatics*. London: Longman.
van Leeuwen, Theo. 2007. Legitimation in discourse and communication. *Discourse & Communication* 1(1). 91–112.
Limberg, Holger. 2009. Impoliteness and threat responses. *Journal of Pragmatics* 41(7). 1376–1394.
Nobata, Chikashi, Joel Tetreault, Achint Thomas, Yashar Mehdad & Yi Chang. 2016. Abusive language detection in online user content. In *Proceedings of the 25th international conference on world wide web, WWW*, 145–153.
Penman, Robyn. 1990. Facework and politeness: Multiple goals in courtroom discourse. *Journal of Language and Social Psychology* 9(1–2). 15–38.
United Nations. 2012. The Rabat Plan of Action on the Prohibition of Advocacy of National, Racial or Religious Hatred that Constitutes Incitement to Discrimination, Hostility or Violence. https://www.ohchr.org (accessed 14 May 2022).

Ruzaitė, Jūratė. 2021. How do haters hate? Verbal aggression in Lithuanian online comments. In Innocent Chiluwa (ed.), *Discourse and conflict: analysing text and talk of conflict, hate and peacebuilding*, 115–145. London: Palgrave Macmillan.

Salminen, Joni, Maximilian Hopf, Shammur A. Chowdhury, Soon-gyo Jung, Hind Almerekhi & Bernard J. Jansen. Developing an online hate classifier for multiple social media platforms. *Human-centric Computing and Information Sciences* 10(1). 1–34.

Shum, Winnie & Cynthia Lee. 2013. (Im)politeness and disagreement in two Hong Kong internet discussion forums. *Journal of Pragmatics* 50(1). 52–83.

Taylor, Charlotte. 2015. Beyond sarcasm: The metalanguage and structures of mock politeness. *Journal of Pragmatics* 87(1). 127–141.

Terkourafi, Marina. 2002. Politeness and formulaicity: Evidence from Cypriot Greek. *Journal of Greek Linguistics* 3(1). 179–201.

Tselika, Aikaterini. 2015. Irony as an impoliteness tool: An exploration of irony's intentionality, cancellability and strength. *Athens Journal of Philology* 2(2). 89–108.

Vasilaki, Maria. 2014. Name-calling in Greek YouTube comments. In Carolina Pérez-Arredondo, Margarita Calderón-López, Hilda Hidalgo-Avilés & David Pask-Hughes (eds.), *Papers from the 9th Lancaster University postgraduate conference in linguistics & language teaching*, 90–110. Lancaster: Lancaster University.

Walters, Mark Austin. 2013. Conceptualizing 'hostility' for hate crime law: Minding 'the minutiae' when interpreting Section 28(1)(a) of the Crime and Disorder Act 1998. *Oxford Journal of Legal Studies* 34(1). 47–74.

Fabienne Baider
Chapter 7
Covert Islamophobia and anti-Semitism via conspiracy theory

1 Introduction

Research on migration discourse in the press, in political speeches, and in online comments has emphasised the negative stereotyping of migrants and has noted the repeated use of the THREAT metaphor (Stephan et al. 2016; Musolff 2007; Baider & Kopytowska 2017). Indeed, as Wohlfeld (2014: 63) concluded, migration has become a national issue that is instrumentalised politically and locally. In some countries, it is even considered a security threat or challenge (Wohlfeld 2014: 63). It is not surprising, therefore, that studies on hate speech often focus on migration, since the phenomenon yields a rich corpus of offensive and unacceptable speech (Fortuna & Nunes 2018). However, tagging comments as hate speech is by no means an easy task: the utterance's illocutionary force (the intent of the speaker) can be difficult to prove, while any assessment of perlocutionary effects must consider psychological and sociological contexts. To make matters worse, covert xenophobic statements, i.e. innuendos that spread stereotypes and prejudices surreptitiously, have a similar illocutionary force to those of overt hate speech (the intention to cause harm or to call for violence) and potentially have a similar perlocutionary effect. Given the power of covert hate speech, and the challenges it raises for legal reasoning, I write this chapter with two purposes. First, the chapter proposes a framework to analyse covert hate speech, based on Langton's studies (2012, 2018) focusing on hate speech's practical and epistemic authority. Second, this chapter assesses the limits and potential of such a framework by analysing one specific discursive strategy of covert hate speech: anti-Semitic and Islamophobic conspiracy theories. Our discussion is based on the data collected in an EU project focused on the detection of hate speech online.[1]

---

[1] This article is based on data collected within IMsyPP, Innovative Monitoring Systems and Prevention Policies of Online Hate Speech, co-funded by the European Union Rights, Equality and Citizenship Programme (2014–2020). The project IMSyPP, grant No. 875263, is available at http://imsypp.ijs.si/. The European Commission's support for the production of this publication does not constitute an endorsement of the contents, which reflect the views of the authors alone, and

Fabienne Baider, University of Cyprus

https://doi.org/10.1515/9783110789157-007

2 Defining overt and covert hate speech

Although several recent studies address the phenomenon of covert hate speech (Ben-David & Fernández 2016; Bhat & Klein 2020; Baider 2020, 2022a; Gelber 2021), there is no precise definition of the term—most attempts to define covert hate speech are restricted to adding qualifiers, such as "*soft* hate speech" (Baider et al. 2017; our emphasis). In this section, I propose a definition that lies at the interface of law and linguistic pragmatics and which draws on Langton's work (2012, 2018). Before we proceed, I must first summarise the major definitions of hate speech.

2.1 Hate speech definitions

Definitions are fundamental to any legal action related to hate speech. While there is no universally accepted definition of the phenomenon, diverse supra-national documents (and their national transpositions) offer descriptions of hate speech and the consequent parameters for its prohibition (see Chapter 9 in this volume). Such a discordant legal setting has social consequences. In a prior study (Baider 2022a), I examined important definitions of hate speech in international law, finding that:

1) The International Convention for the Elimination of all forms of Racial Discrimination, agreed by the United Nations General Assembly in 1965, describes hate speech as promoting racial hatred and discrimination, based on "ideas or theories of superiority of one race or group of persons of one colour or ethnic origin".
2) The International Covenant on Civil and Political Rights (ICCPR), decided at the United Nations General Assembly in 1966, prohibited "Any advocacy of national, racial or religious hatred that constitutes incitement to discrimination, hostility or violence".

With regard to EU law, the Council Framework Decision of 2008 confines hate speech to:

3) [All] conduct publicly inciting to violence or hatred directed against a group of persons or a member of such a group defined by reference to race, colour, religion, descent or national or ethnic origin.

---

the Commission cannot be held responsible for any use which may be made of the information contained therein.

Together these laws identify two core components of hate speech: the speaker's *intention* to inflict harm and to *incite* discrimination, hostility or violence towards or against the target groups. The ICCPR defines incitement as: "statements that create an imminent risk of discrimination, hostility or violence against persons belonging to specific categories". It is a concept that is key to the prosecution of hate speech crimes. For example, if a person incites [someone] to act as described above, they can be judged guilty even if no actual offence against a national law has been committed. Once incitement is proven, the offence can be considered hate speech. Incitement, therefore, presupposes the *intention* of harming, since hate speech *per se* would be labelled an inchoate offence. Hate speech is, therefore, not only an expressive speech act, in the sense of "venting the hatred that is boiling inside" (Waldron 2012: 2), it also has a *potential* or *attested performativity*, which Matsuda describes as "words that wound" (Matsuda et al. 1993; Waldron 2012; Langton 2018).

To illustrate the complexity of the way in which the speech act expresses hate, let us consider the following examples. Each fulfils the above-mentioned basic criteria of hate speech, i.e. the intention to inflict harm and the call for/incitement to violence or hatred against a specific and vulnerable community:

(1) There is only one solution [. . .] to burn every XXX. To eliminate them, no one should be left alive. (Baider & Bobori 2019: 104)

(2) You have to combat them (migrants) because they are spies, thieves, swindlers, rapists, criminals, and bandits. (Baider & Kopytowska 2017: 201)

The first quotation is a call for violence: it gives a direct order to burn and eliminate a community. In the second, the word *fight* incites the reader to violence, and the words *spies, thieves, swindlers, rapists, criminals,* and *bandits*, which all share the same semantic field, goad the audience into changing their minds about a specific community. Both messages could be legally prosecuted, because they urge the reader to commit a crime; it is irrelevant whether these orders are followed or not. Moreover, while one can argue that the act of giving such orders reveals the *intention* to harm, how can we *prove* that these orders entail the *intent* to *incite* violence? Earlier research has pointed out the fact that an intention involves a plan of action (Culpeper 2021). However, with online speech a plan of action and an intention can be difficult to prove, since the author can easily argue that they wrote on the spur of the moment and in the heat of the discussion, without thinking—without premeditation and, hence, without intention (Oz et al. 2018). We must conclude, therefore, that a hate speech act can be performative, without an intention to harm. We concur with Langton (2018) that the authority of hate speech, i.e. the hate-advocating speaker's authority, must be considered when prosecuting or analysing speech acts.

2.2 Authority of hate speech acts

Langton (2018: 126–128) suggested analysing the performativity of hate speech at two levels: epistemic authority and practical authority.

2.2.1 The epistemic authority of hate speech

In examples (1) and (2) in section 1.1, each quotation encourages a change of mind among its readers, i.e. each message asserts epistemic authority, since each message can be characterised by its propagandising effect. In this case, the message is assessed for its success in spreading prejudiced statements and negative stereotypes about an outgroup. The persuasive effect of such speech affects the targets of these statements, i.e. the outgroup who are likely to be frightened and anticipate imminent repercussions. Indeed, such statements spread invented narratives or false assumptions that can incite (violent) action. In (2), the reader can see how the words *spies, thieves, swindlers, rapists, criminals,* and *bandits* propagandise negative stereotypes and feed prejudice. Therefore, the statement would count as hate speech.

2.2.2 The practical authority of hate speech

The quotations above also encourage a change in behaviour and seek to incite action, since they call for an assault on a group or individual. This incitement constitutes the message's directive authority or practical authority. This practical authority is found in speech acts encouraging violent actions against specific targets. In (1), there is an explicit call for violence ("to burn every XXX and eliminate them; no one should be left alive"). In (2), there is an indirect order ("you must combat them"). Both speech acts, therefore, assert practical authority. Whether such statements result in actual discriminatory behaviour depends on the recipients.

The epistemic and practical authorities asserted in both messages have incitement to discrimination—and consequent hatred, hostility, or even violence—against the target groups as their sole aim, an aim pursued by recirculating harmful stereotypes and prejudices. These speech acts, which represent the target groups as threatening or dangerous, can trigger strong emotions in their victims, such as fear for one's safety or survival (Stephan et al. 2016).

2.2.3 Legal application of epistemic and practical authority criteria

For examples of hate speech convictions associated with epistemic authority, we cite the court cases *Erbakan v. Turkey* (Judgment of 6th July 2006, § 56) and *Belkacem v. Belgium* (27th June 2017).[2] In the former, the court considered it necessary to sanction the forms of expression used because they spread and promoted hatred based on intolerance. In the case of Belkacem, the defendant, interestingly, argued that "[he] had never *intended* to *incite* others to hatred, violence or discrimination but had simply sought to *propagate* [his] ideas and opinions" (our italics), i.e. explicitly using the two notions of epistemic authority and practical authority. The court declared the argument inadmissible because the defendant called on viewers "to overpower non-Muslims, teach them a lesson and fight them". Thus, the court acknowledged the practical authority of Belkacem's speech act. He was subsequently convicted of hate speech— "inciting discrimination, hatred and violence" towards non-Muslims. In this case, the defendant acknowledged the epistemic authority of his speech but denied its practical authority. It can be argued that the propagation of such threats satisfies the requisite conditions for labelling the statements as hate speech.

Scholars have also distinguished two types of consequences (Langton 2012: 76–80, 86–89; Maitra & McGowan 2007: 62) for each category (epistemic and practical): *consequential harm* (related to practical authority), which refers to changes in listeners' or readers' behaviour, including speech-incited acts of violence; *constitutive harm* (related to epistemic authority), which refers to changes in readers' or hearers' attitudes. Each type of consequence will be discussed at length, with examples, in section 4.

2.3 Covert hate speech

Gelber (2021: 393) argues that hate speech has been regulated in the European Union because it prohibits "equal participation in democratic deliberation". However, there could be a discrepancy between the legal understanding of the term hate speech and the multiple forms—including many covert forms—it can take. Recent research has focused on the illocutionary force and perlocutionary effect of offensive and abusive speech motivated by prejudice but which does not openly express hatred nor call for violence (Ben-David & Fernández 2016; Bhat & O Klein 2020; Baider 2020; Gelber 2021).

---

2 http://www.echr.coe.int/Documents/FS_Hate_speech_ENG.pdf

2.3.1 Covert hate speech definition

More often than not, no restrictions can be imposed on offensive, insulting or abusive speech, because this would impinge "in an overly broad and unacceptable way on freedom of speech, which is a core liberal and democratic value and essential to democratic governance" (Gelber 2021: 393). However, speech acts that cause harm by influencing an audience through the expression of discriminatory beliefs, thereby changing attitudes and behaviour towards a specific community, may not outwardly appear hateful. To qualify as hate speech, language does not have to contain overt insults or explicit calls for violence, nor reproduce racist biases through conventional references such as explicit stereotypes (Hill 2008: 41). Language can communicate calls for violence through implicatures. That is to say, through implied meanings, beyond what is explicitly stated: "implicatures [. . .] can arise in two ways: via presumptions concerning rational communicative behaviour or via certain linguistic convention" (Carston & Hall 2012: 49). In fact, racism and/or xenophobia today is most often conveyed through these less direct or more subtle forms of discrimination (Haijan & Domingo-Ferrer 2013; Taguieff 2005). Since the implementation of the 2016 Code of Conduct in the European Union cyberspace, we have increasingly seen indirect and semiotic calls for violence and hatred online. Statements whose meaning is implied but unmistakable, and which spread prejudice, possibly leading to violence, constitute *covert* hate speech. Thus, we suggest the following definition for covert hate speech:

> Any statement that covertly asserts epistemic authority by clandestinely spreading discriminatory beliefs and encouraging harm against a specific community or an individual belonging to that community, or any statement that implicitly asserts practical authority by encouraging violent action against a specific target or by encouraging constitutive harm against a specific community or an individual belonging to that community.

Covert hate speech can take many forms (Ben-David & Fernández 2016; Bhat & Klein 2020) and can exhibit different degrees of covertness (Ben-David & Fernández 2016: 1170–1171). For example, the use of deliberate misspellings and otherwise coded messages (Magu & Jiebo 2021);[3] the use of cyphers known by the initiated—such as "juices" or "Zionists" for Jews (Warner & Hirschberg 2012); the use of specific symbols such as avatars or pseudonyms covertly referring to Nazism, such as "88" for "Heil Hitler"; sharing links to white supremacist textual or video content online; "liking" hate speech comments on social media; spreading racial conspiracy theories, predominantly anti-Semitic ones (Baider 2022a, 2022b).

---

3 The authors call covert hate speech *euphemism*.

2.3.2 Covert hate speech prosecution

The question is therefore: can covert hate speech be prosecuted while still respecting freedom of expression? Freedom of expression, a fundamental principle in a democratic society, demands respect for differences of opinion, tolerance of difference, and broad-mindedness. Hence, expressing ideas or opinions regarded as offensive, shocking or disturbing must be allowed. Any restriction to freedom of expression should "be proportionate to the legitimate aim pursued" (*Handyside v. the United Kingdom*, judgement of 7th December 1976, § 49).

Freedom of expression can only be restricted in those cases in which the form of expression has spread, promoted, justified or incited hatred based on intolerance (*Erbakan v. Turkey*, judgment of 6th July 2006, § 56). We might conclude, therefore, that covert hate speech that spreads, promotes or incites hatred should also be prosecuted. For this to happen, it must somehow be proven that covert hate speech statements infer the same misleading speech act or that the impetus that can lead to (violent) action in much the same way as openly expressed statements. Consequently, covert hate speech must be shown to assert the same type of epistemic authority or practical authority as does overt hate speech.

2.3.3 Covert hate speech and migration discourse

As mentioned earlier, covert anti-immigration discourse is expressed using several discursive strategies, which include metaphors, irony, sarcasm, and innuendo—that is, allusive or oblique remarks or hints, typically suggestive or slyly disparaging ones.[4] To elucidate these covert strategies, we establish both their epistemic and practical authority, and we use a hybrid legal-linguistic approach. Some recent studies on hate speech have adopted this combination of approaches, e.g. Guillén-Nieto's work on defamation (2020), workplace sexual harassment (2022) and hate speech (2023) (see also Chapter 8 in this volume).

---

4 Sarcasm in covert hate speech has been analysed in Baider and Romain (in press). Interestingly, Facebook accepts some offensive but humorous content (Fortuna & Nunes 2018: 5).

The statement below does not directly call for violence:

(3) WE REAL POLES WILL NEVER BE FOR RECEIVING THOSE MURDERERS, THIS XXX SECT, IT'S A THREAT TO MODERN COUNTRIES IN EUROPE AND THE WORLD! POLAND NOT FOR MIGRANTS !!!!! (Baider & Kopytowska 2017: 199)[5]

This is not a direct call or incitement to violence, but it is an obvious incitement to discrimination, hatred, and hostility; hence, there is practical authority. Moreover, calling migrants murderers implies they are a physical threat to the host population, a threat which is explicitly stated in the same comment. Psychologists who study emotions and conflict have found that a core component of hate speech related to migration is that the speaker presupposes the target to be dangerous to the Self (Stephan et al. 2016). The reasoning goes: if migrants are considered dangerous, then people have the right to defend themselves against them. In this sense, then, the message covertly incites violence, anger, fear and hatred, and can be the origin of causal harm, i.e. practical authority (see Chapter 2 in this volume, on the concept of danger in relation to migration). We can conclude that the writer's intention and the potential consequences of the implicit message (covert hate speech) are similar to those of explicit messages or overt hate speech.

In the same way, the quotation below calls indirectly for hatred and/or violence:

(4) Islam out—Protect the British people

This message was inscribed on a British National Party (BNP) poster displayed in the window of a British home, accompanied by a photo of the World Trade Centre towers in flames, and an Islamic crescent defaced by a large X, signifying "No Islam." The photo of the World Trade Centre reiterates the order "protect the British people", while the crossed-out Islamic crescent metonymically repeats the direct commands (directive speech acts) to bar entry to Muslims while urging the prohibition of Islam in the UK.

The message addresses both communities: the outgroup (Muslims) and the ingroup (white, culturally Christian British people), thus creating or reinforcing social polarisation. The reason given for the proposed ban is that Islam poses a threat to British nationals and to British culture. The outgroup is stigmatised as a terrorist group, while ingroup members are urged to protect themselves and their community.

---

5 Throughout this paper, we have retained the actual phrasing, grammar, spelling, and diacritics used in all comments; (-*-*-*-) indicates that what follows is an answer to a previous comment. Here, we have kept the original capital letters.

Indirectly, this statement calls for violence against the outgroup, which is depicted as extremely dangerous. We can also infer that the outgroup will feel offended and frightened about possible repercussions of such a poster. Combined, the text and the images form a semiotic sign possessing both epistemic authority (telling people what to think) and practical authority (telling people what to do). Because of the fear provoked in the ingroup, ingroup and outgroup may come to mistrust each other (acting on the persuasive effect of hate speech); some ingroup members may decide to attack the targeted community, an example of how hate speech can change behaviour even to the point of violence. The person who posted the picture in his window was subsequently convicted of aggravated hostility towards a religious group, i.e. convicted of a hate speech crime. After an appeal, referring to the right to freedom of expression, the court upheld the conviction. The court found that the speaker's communicative intent (the illocutionary force of the statement) was to incite hatred, hostility or violence against a religious group, because it associated the outgroup with a serious act of terrorism.

The addressees of examples (3) and (4) can draw inferences based on "common ground" (Kecskes & Zhang 2009: 346–351). With hate speech, this common ground is made up of shared prejudices and stereotypes, which examples (3) and (4) reinforce and recirculate. In particular, the presence of migrants is presented as a threat to the nation's well-being in all its facets, physically, economically and culturally (Stephan et al. 2016; Baider & Kopytowska 2017).

In the following section we will focus on a specific phenomenon of covert hate speech, one not often studied but nonetheless pervasive in the data: conspiracy theories.

3 Covert hate speech and conspiracy theories

We have discussed and defined the epistemic authority and practical authority of covert hate speech. In this section we focus on conspiracy theory as a strategy of covert hate speech, one which is used to explain the migration phenomenon.

3.1 Offline covert anti-Semitism and conspiracy theory

The most common discursive strategies used to demonise specific Others and spread prejudices, as Wodak has observed, are: "scapegoating, blaming the victim, victim-perpetrator reversal, trivialisation and denial" (2015: 206). A recent case of scapegoating prosecuted as hate speech is that of Cassandre Fristot in

France, who was convicted in October 2021 (Baider 2022a).⁶ This former member of the *Rassemblement national*, Marine Le Pen's party, was waving a cryptic sign, bearing the words in red all-caps *MAIS QUI?* 'But who?' (see Figure 7.1) during an anti-vaccine demonstration in August 2021.

Figure 7.1: Fristot's sign displayed at a protest in August 2021.

The question "But who?" was decorated with devil's horns crowning the word *QUI*, and was the focal-point of the sign. Surrounding the question were the mostly Jewish-sounding names of prominent French politicians (Macron, Buzyn, Fabius, Attal, B.H.L) and international financiers (Drahi, Soros). There were no calls for violence nor direct insults, but rather a series of innuendos that only the "enlightened" would correctly understand. The rhetorical question "Who?" had become a rallying point hashtag, acting as "an anti-Semitic dog-whistle hinting at shadowy Jewish grey eminences behind the Covid pandemic and the vaccination campaigns" (Audureau 2021, cited in Becker 2022: 10). Fristot's cryptic sign was indeed related to a conspiracy theory that interpreted the almost mandatory COVID-19 vaccines and the restrictions on the freedom of movement of unvaccinated persons as proof of a conspiracy named the New World Order (see the next section). Fristot was arrested for hate speech against the Jewish community in August 2021 and subsequently, in October 2021, was given a six-month suspended sentence and ordered to pay a fine, both to the NGOs, who were among the plaintiffs, and to the State.

This example shows the power of conspiracy theories in providing simplistic answers to complex issues (Covid pandemic, vaccinations), as explained in Nicolas (2014: 2). It also shows the power of such theories to polarise social discourse

---

6 The sign, as well as reactions to it, was extensively analysed in Baider (2022).

and interactions by spreading prejudice against a specific community, which here is mainly the Jewish community, but also more broadly the Establishment. I hypothesise that conspiracy theories are, in fact, a covert strategy of hate speech, one that has been mostly ignored in studies of hate speech.

Before we proceed, however, we must address the question: what is a conspiracy theory? Although conspiracy theories have been variously defined, Popper's definition is widely cited:

> It is the view that an explanation of a social phenomenon consists in the discovery of the men or groups who are interested in the occurrence of this phenomenon (sometimes it is a hidden interest which has first to be revealed), and who have planned and conspired to bring it about. (Popper 1994: 306)

Conspiracy theorists identify the groups described in the quote above as the most powerful financiers, economists and business people in the world. This small but highly influential group is thought to engineer all major world events, whether they are economic crises, political coups, or even pandemics. The objective of the conspirators is said to be world domination for their own benefit, by installing a "New World Order" (often abbreviated as "NWO") at the expense of the common good. In New World Order conspiracy theory, most of the key conspirators are said to be Jewish. Our aim here is to observe the propagation of such conspiracies in online debates in order to: 1) identify the keywords used for purposes of algorithmically detecting traces of conspiracy theory and 2) work on the creation of counter-speech to respond to allegations of conspiracy.

3.2 Evaluating online covert hate speech

We worked from a dataset called FRENK[7], which comprises Facebook posts and comments related to mainstream news media, mostly from Great Britain. FRENK covers two topics: migrants/immigration and the LGBT community. We focus here on immigration, as the comments were collected from posts made on Facebook in 2015 when an unprecedented number of migrants arrived in Europe. In a previous project, we had already manually annotated each comment for different types of socially unacceptable discourse (Ljubešić et al. 2019). The dataset is in English (see Table 7.1).

---

7 The acronym FRENK stands for "FRENK – Raziskave Elektronske Nespodobne Komunikacije" (Engl. "Research on Electronic Inappropriate Communication").

Table 7.1: Datasets used for annotations.

DATA				
Number of comments	Number of annotated comments	Source of dataset	Topic	Language
5,873	3,700	Facebook	Migration	English

To analyse rhetorical strategies used to arouse hatred towards migrants, we first had to decide on a set of parameters for annotating the comments, so we turned to earlier research studies and reviewed criteria used in several other domains known for their work on hate speech: data mining (for example, Fortuna & Nunes 2018; Sanguinetti et al. 2018), psychology (Stephan & Stephan 1999), discourse analysis and rhetoric (Wodak 2015; Baider 2019; 2020). We determined a set of criteria to evaluate the rhetoric used, then tested our criteria in a 2-week pilot study:
1. *Argumentative rhetoric* (subdivided into logic/reasoning, statistics, examples, history, conspiracy theory);
2. *Affective rhetoric* (subdivided into insult, acknowledgement of grievances, displays of positive emotions, displays of negative emotions, irony and sarcasm).

Conspiracy theory was, therefore, only one of the parameters we chose to evaluate the argumentative strategies used. Over three months, our Cypriot team analysed 3700 comments and tagged what we considered *argumentative* or *affective* strategies in comments counted as hate speech. Each comment was annotated by eight annotators working in pairs. Most comments were in response to a posting from the BBC captioned: *As part of #myrefugeestory, Mohamed Madi and BBCJamesLongman* [sic] *met Syrian refugees in one of Beirut's hundreds of informal settlements.*

Among the comments labelled as conspiracy theory, most of the discussion threads contain the word "invasion". Sanguinetti et al. (2018) also observed that in Italian Twitter posts words such as *invasore* 'invader', and *invasione* 'invasion', occur frequently in texts labelled in immigration data as hate speech. The names *Soros* and *Rothschild*, and the terms *Israel* and *Jews* were also relevant. These keywords were identified in our earlier research as indicative of conspiracy theories (Baider & Sini 2021; Baider 2022a, 2022b). Soros, for example, is a Hungarian-born businessman and billionaire who supposedly heads a "sinister globalist clique", in the words of Victor Orban (Baider 2022b). Names like Soros function similarly to the *QUI* on Fristot's sign (and in the French hashtag #QUI): they have become rallying points for New World Order conspiracy theorists.

In fact, all these keywords are core to the New World Order conspiracy theory described above,[8] which, in our analysis, is related to Islamophobia and anti-Semitism.

4 The Great Replacement Theory: covert Islamophobia, covert anti-Semitism

4.1 Covert Islamophobia

A sub-theory of the New World Order theory is called the Great Replacement Theory. This theory targets migration, arguing that waves of non-Europeans have been organised to invade the EU. These mass arrivals have as their ultimate aim the replacement of most indigenous Europeans. Immigration is, therefore, seen as a genocidal threat to European people, culture and civilization (Taguieff 2005; Baider 2022a, 2022b). According to this theory, the conflicts in the Middle East—notably those that caused the wave of immigration in 2015, and the Marrakech pact on migration in 2018—were knowingly organised and planned. Behind these events, supposedly, are the occulted forces of the New World Order who have formed a secret alliance (of corrupt politicians, greedy bankers and industrialists without faith or law) to govern the world, while the United Nations is proof of their intention to dissolve all (European) national identities.

In the following section, we focus on the threads within online discussions that give voice to this theory. Our aim is to understand their frequency (question 1) and their impact, via the responses they generate (question 2).

4.1.1 A planned Islamic invasion

Most comments that annotators tagged as conspiracy-theory-related contained the keyword *invasion* and its paradigm (20 or 30 comments). The number preceding

---

8 There were other conspiracist comments involving political argumentation, such as in the quote below about the West fomenting terrorism in the East, but they were few:
    112 What we see emerging is what was suspected and even obvious all along—a proxy war started by, and fought for Western hegemonic ambitions in the region, intentionally feeding the forces of extremism, not fighting them.

each utterance refers to the comment's numeration in our database. Below are the comments we will be analysing in this section.⁹

> 311 -*-*-*- *Invasion forces* usually do comprise of young men of fighting age, dear.
> 351 BBC please report the truth [. . .] This is not a humanitarian crisis, *this is an invasion*!
> 360 the hungarians are correct *it is an invasion*, and it is aided and abbeted by the BBC and Sky news
> 414 It's a *muslim invasion*. [. . .] *it's a planned invasion* and we need to seal the borders and send those that are here back
> 553 [. . .] Muslims led *many invasion* in the past but they no longer need armes to invade Europe. [. . .]
> 599 Seems more *like some kind of invasion*—so many military aged young men in those groups than a real refugee movement.
> 3163 Lawlessness! *Pure invasion*!!
> 3183 a simple *strategy of invasion*.......
> 3379 -*-*-*- *its invasion and planned way and lots manipulation* and that is done before upcoming election in US...
> 3439 -*-*-*- Your so right. This is an *invasion* and they will cause WW3.
> 4010 So, wearing a full hijab and dehumanizing women is now dignity? Only when *invaders* do it though.

The prevalence with which the migration phenomenon is called an "invasion" is no longer surprising: for some time now, studies of migration discourse (Santa Ana 1999; O'Brien 2003; Musolff 2015; Sanguinetti et al. 2018) have revealed how commonly this metaphor is used with reference to migration. The metaphor of invasion is prominent among the classic examples of metaphors about migration.¹⁰ This reconceptualisation of migration as invasion entails a scenario very different from the typical migration scenario (Baider 2017). In (360) and (351) the true reasons for the migration (flight from war) are concealed. The social fact (emergency compelled movement of a population) is renamed and reconceptualised as a planned, deliberate move:

> 351 BBC please report the truth [. . .] This is not a humanitarian crisis, *this is an invasion*!
> 360 the hungarians are correct *it is an invasion*, and it is aided and abetted by the BBC and Sky news

---

9 We avoided numbering the quotes, given that each comment is already numbered.
10 Among other classical metaphors is immigration as a natural disaster as in (501) but it was much less present in our data:
> 501 It needs to stop now *these migrants are flooding* through Hungary turned back Bangladeshis last week . . .. This is *not a refugee crisis it's a migrant invasion*.

The word *invasion* implants a set of assumptions about typical participants and their roles, the storylines and the conventional evaluations (Musolff 2006: 27–28). The vulnerable asylum seekers are now understood as being bloodthirsty soldiers, in (311) and (599), whose actions will trigger a worldwide war (3439):

> 311 -*-*-*- *Invasion forces* usually do comprise of young men of fighting age, dear.
> 599 Seems more *like some kind of invasion*—so many military aged young men in those groups than a real refugee movement.
> 3439 -*-*-*- Your so right. This is an *invasion* and they will cause WW3.

These invading forces, described as Muslims (414), are seen as reminders of the armies of the Ottoman empire (553). Such reframing of a current social issue (immigration) in terms of a historical scenario familiar to many Europeans (the Ottoman invasions, between 1356 and 1565) can revive long-buried or subconscious fears, and/or elicit intensely negative reactions.

> 414 It's a *muslim invasion*. [. . .] *it's a planned invasion* and we need to seal the borders and send those that are here back
> 553 [. . .] Muslims led *many invasion* in the past but they no longer need armes to invade Europe. [. . .]
> 3163 Lawlessness! Pure invasion!

Comment (4010) makes a false claim, i.e. that the "invaders" will apply Sharia law—yet Sharia law applies only in certain Muslim countries, not in others (Baider & Bobori 2019).

> 4010 So, wearing a full hijab and dehumanizing women is now dignity? Only when *invaders* do it though

Fallacies and fallacious reasoning are common online argumentation strategies. Krobová and Zàpotocky (2021) concluded from their analysis of online comments related to migration on Czech Facebook that these rhetorical devices serve first as appeals to fear (epistemic authority), and consequently represent an implicit call for action (practical authority).

Framing migration as an invasion implies as well that readers should *resist* this planned movement, by, for example, closing the borders: "it's a planned invasion and we need to seal the borders" (comment 414), or even potentially take up arms, as in comment (3185), where there is an overt call for violence against migrants: "Crush the invaders. Give them the sword". The noun *sword* evokes the Crusades, with the aim of suggesting the need for a reverse holy war, to resist the attempt by the "invaders" to "Islamise" the West.

> 3122 The Islamization of the West, was a well-planned move.
> 3123-*-*-*- It was implemented in the early 90's in Pakistan.

Most importantly, describing migrant arrivals as an invasion usually implies an organised movement, rather than a painful move compelled by a national disaster. Comment (3379) outright declares that these arrivals have been planned: "its invasion and planned way and lots manipulation". Comment (414) does the same: "it's a planned invasion and we need to seal the borders".

The words in (3122) and (3123) above, "well-planned" and "implemented", or the phrase in (3183), "A simple strategy of invasion", infer that someone or some group behind the scenes is organising a mass movement into Europe. The behind-the-scenes organiser could be the Establishment, a major culprit in conspiracy theories, as represented by the mainstream media: "it [invasion] is aided and abetted by the BBC and Sky news" (3501). In fact, many conspiracy theorists believe that the Jewish-controlled media support the Establishment in creating a New World Order that will lead to the erasure and "replacement" of Europeans.

4.2 Anti-Semitism

As mentioned earlier, the New World Order theory implies the wrongdoings of Jewish people, described as being at the core of the Establishment behind the New World Order, actively promoting the Great Replacement. Indeed, Jews are most often named as the leading agents behind the migrant invasion scenario.

4.2.1 Spreading doubt to suggest a scapegoat

The aim of conspiracy theorists is to spread doubt about facts or rational explanations, and to construct links—for our purposes here, links related to migration—in order to further blight the truth with distrust. In our data, we identified three main strategies used to spread doubt: asking rhetorical questions, making baseless innuendos with qualifiers, and using insinuatory ellipses.

El Sherief et al. (2021) have observed that rhetorical questions are routinely used in covert hate speech. In their data, rhetorical questions were most often used to convey negative stereotypes associated with a specific community. In our data, even if the word *invasion* or its paradigm (*invade, invader*) is absent, the use of question marks, such as in comments (3096) and (3153), suggests the author's general doubt over refugees' true motives, stated explicitly in comment (600).

> 3096 "Why are they migrating???"
> 3153 "Are we being invaded?! . . ."

600 Also, does it seem a bit odd that hundreds of thousands of migrants are coming at the same time? Was there an announcement made in all the refugee camps to leave at once? It appears to be a little bit of a conspiracy.

Several posts use rhetorical questions that seem to imply a conspiracy, and these lead the reader to pause. Rhetorical questions do not expect an answer; the speaker already knows the answer. This was the case with the *MAIS QUI?* on Fristot's sign discussed above. Fristot was not *asking* a question, but inviting the public to *read* her (anti-Semitic) implicature. The strategy functions as an *interactional* device—in our example, there is an intent to spur conversation among those who deem themselves "enlightened" about the conspiratorial "truth" behind the influx of refugees into Europe in 2015.

In the same indirect way, to imply a conspiracy, qualifiers are used to emphasise the peculiar suddenness of the event (mass migration) and its consequent resemblance to an organised movement. Comment (600) includes qualifiers ("a bit", "a little bit") expressing modality. If the functions of modality are in principle numerous, here the qualifiers are used to increase the chances that reluctant readers will pause to consider the argument and then eventually concur. It also invites readers to invert the meaning of "*a bit* odd" and "*a little bit* of a conspiracy" into "*very* odd" and "*very much like* a conspiracy". These statements negotiate "interactional meanings in a text, assisting the writer (or speaker) in expressing a viewpoint and engaging with readers as members of a particular community" (Hyland 2005: 37). Reader engagement in this case enhances the chance of persuading them to consider an "alternative truth".

Using ellipses is another strategy for conveying innuendo, common in this dataset. Comments (601), (749), (3153) and (3183) end in ellipses, inviting readers to question their own thoughts on the issue, by implying that something important has been left unsaid.

601 -*-*-*- First of all these people are not migrants . . ..they are missionaries of you know who . . . ! [. . .]. So, decide it for yourself . . .
749 Why not neighbouring Israel . . .?
3153 Are we being invaded?!.
3183 a simple strategy of invasion. . . . . . .

Such usages not only spread doubt, they also point to a scapegoat, which here is Israel.

4.2.2 Covert anti-Semitism

The supposed role of Jewish people in driving this mass migration is central to many comments in the dataset, as we can see in the exchanges quoted below. Purported instigators of the migration movements are mentioned early in the debate: the CIA, supposedly an arm of the NWO, and George Soros, supposed head of a globalist clique of financiers. Commenter (88) attempts to halt the verbal violence of other commenters with a rational argument: it is the wars in the Middle East, initiated by Western powers, that explain the surge in refugees.

> 88 -*-*-*- Wars in the Middle East Have Created More Terrorists!! [. . .] If we want a better world then there shouldn't be selective humanity and sympathy, Emphasis should be on as you stated 'NO HUMAN SHOULD SUFFER.'
> 90 -*-*-*- you are absolutely right regarding the cause for all of this misery . . . *The CIA should have stayed the hell out* of everyone else's business . . . *But behind the CIA is the new world order and George Soros . . . People are waking up to the tru*th . . . And they got everyone fighting everyone else while they steal everybody's freedom

Commenter (90) agrees that wars have caused the sharp rise in migration, but then argues that the wars are the result of a conspiracy by actors such as the CIA and Soros—agents of the New World Order.[11] Therefore, when (601) answers the rhetorical question posed by commenter (600) with the cryptic "you know who", we understand from comment (90) that migrants are "missionaries" of the CIA and of Soros:

> 601 -*-*-*- First of all these people are not migrants . . . ..they are *missionaries of you know who* . . . ! [. . .] So, decide it for yourself . . . *are they refugees or part of a new Jihad planned by ISIS*. (The best way to infiltrate Europe)

The second part of the comment mentions ISIS, which seems irrelevant to the conspiracy theory, Soros or the CIA. However, a subsequent comment clarifies the reasoning at work: ISIS is the creation of the USA, NATO, Israel and Saudi Arabia; ergo, we can infer that they also created the migration waves:

> 619 Obama, NATO, Israel & Saudis start this mess Funding the Syrian Rebels aka Al Qaeda / ISIS with Money & Weapons trying to overthrow Bashar al-Assad and kill all the Syrian Christians.

---

[11] Anti-Israel comments also sometimes refer to Israel's low refugee intake—a country that "is supposed to be the most humane in the region?" (comment 3335).

Later in the thread, other commenters try to debunk this theory. Comment (3353) argues that spreading such rumours actually helps ISIS in their aim to weaken the West.

> 3353 -*-*-*- We know perfectly well that supporters of ISIS continually try to claim that USA/ Israel / UK or the West in general, created and supports ISIS. Of course, no real evidence is given to support these claims. That is not necessary, as the claims are only devised to make gullible people hate the West. Which is an aim of ISIS.

In the next section, however, we will see that the argument made in comment (3353) was ineffective.

4.2.3 Overt anti-Semitism

Indeed, in response to suggestions that someone/something other than Israel is responsible for the migration wave, commenters (3672) and (3763) insist that Israel was instrumental in creating ISIS and the mass migration:

> 3672 *Isis is now proved to be a plan of Israel financed.* Waiting for the day when Israel is destructed for everlasting peace in the world.
> 3763 I *still think it's a conspiracy of Israel IL.* Why they never talked about attacking Israel.

The argument here is that the fact that ISIS has not attacked Israel is evidence of Israel's involvement in the creation of ISIS—ignoring that fact that most European countries have also been spared ISIS' attacks, and that Israel's stringent security measures might also explain the lack of attacks. As we can see, the anti-Semitic comments become more brazen as the thread continues; comment (90) only mentions Soros and the New World Order, while (3672) and (3763) name Israel explicitly.

Commenters (4996) and (4999) then openly identify the "Rothschilds Mafia", who they accuse of conspiring to destroy Europe and the EU by flooding Europe with refugees.

> 4996 -*-*-*- Listen up people! EU is *being ransacked by Rothschilds Mafia*, [. . .], Brussels will be smashed even further, Germany will be along with Spain, France and others, refugee migration is manufactured to destroy Europe by *none other than George Soros (Rothschild Mafia)* wakey wakey people, the 1 thing that these *maniacs* hate is solidarity and community spirit! THINK

The intensity of the emotion—whether of anger, excitement or fear, as revealed with the capitalisation, triple exclamation marks—is heightened by spirited descriptions of destruction and devastation: the EU being *ransacked*, Brussels being *smashed*, Europe being *destroyed* by *maniacs*. The targeted group (migrants) is depicted as a serious threat to Europe, because they are pawns in a geopolitical game (the New World Order). Commenter (4996) urges their audience to react against the world leaders' "manipulation", or at least to think about this theory (comment 4996) and to read between the lines. Using capital letters for THINK emphasises the urgency of the command to change one's mind. Conspiracy theorists often present themselves as "enlightened" and smarter than others, believing that only they can see through the many deceptions to the hidden truth. The names CIA, Rothschild, and Soros are used for their symbolic capital (Bourdieu 1982); i.e. they are a recognisable token among those who subscribe to this conspiracy theory.

To support their claims about the Rothschild family, the only evidence presented by the conspiracists is the fact that none of the countries at war with the west, nor any of those deemed a threat to the west, are home to a Rothschild bank:

> 4999-*-*-*- [. . .] it doesn't matter who you vote for because when it comes to geopolitics neither party leader has any say in the situation what so ever, look at the countries that are deemed a threat or who are at war with the West, they all have one very important thing in common! NONE OF THEM HAVE AN OPERATING *ROTHSCHILD BANK IN THEIR COUNTRY*!!! (our emphasis)

In sum, the illocutionary force of a conspiracist statement serves to assert epistemic authority, i.e. it is an attempt to change people's minds, to convert them. In this case, the recipients of the message are meant to believe that a specific group of people, who happen to have Jewish names, are behind all the wars that have led to mass migratory flows into Europe. This epistemic authority is conveyed as follows:
1. Make other people believe that something is not right, nothing is as it seems, and nobody is what they claim to be; this serves to *create doubt* about the presentation of important events;
2. Convince people of an *alternative truth that involves the Jewish community*: this migration did not happen by chance because it was organised by Jewish personalities, such as Soros, who represent the CIA, the USA and Israel.

Sometimes recipients overtly acknowledge the illocutionary force of innuendos delivered on social media. Note the way commenter (1498) ironically underlines how quickly conspiracy theories flood the conversation whenever immigration is mentioned or discussed: "-*-*- *'Love the fact somewhere along the line, the anti-accepting-refugees lot bring up hidden collusions or '"conspiracy theories"'*. This is exactly what this chapter has tried to demonstrate.

5 Concluding remarks

Overt hate speech is immediately recognised as socially unacceptable and reprehensible. However, as Langton argued (2012: 73), we should not assume that the worst terms of abuse constitute hate speech nor that speech that is not plainly abusive could not constitute hate speech (cf. Baider 2020). In this chapter, we have discussed how conspiracy theories, and more precisely the New World Order and Great Replacement theories, serve as a master narrative for encouraging prejudices against groups of people who have legally protected characteristics. We analysed these conspiracies as a specific covert strategy for building anti-Semitic and Islamophobic sentiment and narratives, which entails similar epistemic and practical authorities as overt hate speech strategies. However, covert hate speech, especially in such forms, can be more persuasive and dangerous than overt hate speech. Indeed, concerning the perlocutionary effect of covert hate speech, inferences triggered by conspiracy theories may be more insidiously effective than overt hate speech in arousing resentment, fear, anger and contempt—certainly the most prevalent emotions connected with hatred and consequent discriminatory violence. However, belief in these conspiracies serves as a common ground among far-right groups and their leaders (cf. Baider 2022b).[12] Orban and Salvini openly refer to the New World Order theory to explain the 2015 migration wave (Baider 2022b). On the societal level, Plenta (2020: 516) argues that conspiracy theories, precisely in their function as common ground, can serve to strengthen community ties among far-righters. More importantly, conspiracy theories can also become an emergent common ground for the uninitiated in online discussions, where they may find themselves persuaded by such theories (Baider 2020, 2017). Social media has proven an extremely effective propagandising tool for conspiracy theories, for this reason. We argue, therefore, that conspiracy theory should be understood as a covert hate speech strategy in any attempts to identify and prosecute online hate speech.

---

[12] This is also true to a lesser degree to far-left partisans.

References

Baider, Fabienne & Lorella Sini. 2021. Le Complotisme transnational et le discours de haine: Le cas de Chypre et de l'Italie. *Mots: Les langages du politique* 125. 15–34.

Baider, Fabienne & Monica Kopytowska. 2017. Conceptualising the other: Online discourses on the current refugee crisis in Cyprus and Poland. *Lodz Papers in Pragmatics* 13(2). 203–233.

Baider, Fabienne. 2017. Thinking globally, acting locally: Mainstream supremacist concepts within a local socio-historical context. *Journal of Aggression Language and Conflict* 5(2). 178–204.

Baider, Fabienne & Anna Bobori. 2019. Mitigating the frame 'sexual threat' migration discourse online. In Darja Fisher & Philippa Smith (eds), *The dark side of digital platforms*, 86–113. Ljubljana: Ljubljana University Press.

Baider, Fabienne. 2020. Pragmatics lost? Overview, synthesis and proposition in defining online hate speech. *Pragmatics and Society* 11(2): 196–218.

Baider, Fabienne. 2022a. Covert hate speech, conspiracy theory and anti-Semitism: Linguistic analysis versus legal judgement. *International Journal for the Semiotics of Law* 35. 1–25.

Baider, Fabienne. 2022b. The dynamic model of meaning approach: Analysing common platform development of conspiracy theory in far-right discourse. In Istvan Kecskes (ed.), *Handbook of intercultural pragmatics*, 711–738. Cambridge: Cambridge University Press.

Becker, Matthias J., Laura Ascone, Matthew Bolton, Alexis Chapelan, Jan Krasni, Karolina Placzynta, Marcus Scheiber, Hagen Troschke & Chloe Vincent. 2022. *Decoding anti-Semitism: An AI-driven study on hate speech and imagery online. Discourse report* 3. Berlin: Technische Universität Berlin.

Bhat, Prashanth & Ofra Klein. 2020. Covert hate speech: White nationalists and dog whistle communication on Twitter. In Gwen Bouvier & Judith E. Rosenbaum (eds.), In *Twitter, the public sphere, and the chaos of online deliberation*, 51–172. London: Palgrave Macmillan.

Ben-David, Anat & Arianda Matamoros Fernández. 2016. Hate speech and covert discrimination on social media: Monitoring the Facebook pages of extreme-right political parties in Spain. *International Journal of Communication* 10. 1167–1193.

Brown, Alexander. 2015. *Hate speech law: A philosophical examination*. London: Routledge.

Carston, Robyn & Alison Hall. 2012. Implicature et explicature. In Hans-Jörg Schmid (ed.), *Cognitive Pragmatics*, 47–84. Berlin/Boston: De Gruyter Mouton.

El Sherief, Mai, Caleb Ziems, David Muchlinski, Vaishnavi Anupindi, Jordyn Seybolt, Munmun De Choudhury & Diyi Yang. 2021. Latent hatred: A benchmark for understanding implicit hate speech. *Proceedings of the 2021 Conference on Empirical Methods in Natural Language Processing*, 345–363. Association for Computational Linguistics.

Gelber, Katharine. 2021. Differentiating hate speech: A systemic discrimination approach. *Critical Review of International Social and Political Philosophy* 24(4). 393–414.

Guillén-Nieto, Victoria. 2020. Defamation as a language crime: A sociopragmatic approach to defamation cases in the high courts of justice of Spain. *International Journal of Language and Law* 9. 1–22.

Guillén-Nieto, Victoria. 2022. "What else can you do to pass . . .?": A pragmatics-based approach to quid pro quo sexual harassment. In Janet Giltrow, Frances Olsen & Donato Mancini (eds.), *Legal meanings: The making and use of meaning in legal reasoning*, 31–55. Berlin/Boston: De Gruyter Mouton.

Guillén-Nieto, Victoria. 2023. *Hate speech: Linguistic perspectives*. Berlin/Boston: De Gruyter Mouton.

Fortuna, Paula & Sergio Nunes. 2018. A survey on automatic detection of hate speech in text. *ACM Computing Surveys (CSUR)* 51(4).1–30.

Hill, Jane. 2008 *The everyday language of white racism*. New York: Wiley-Blackwell.
Hyland, Ken. 2005. *Metadiscourse: Exploring interaction in writing*. London: Bloomsbury.
Kecskes, Istvan & Fenghui Zhang. 2009. Activating, seeking and creating common ground: A socio-cognitive approach. *Pragmatics and Cognition* 17(2). 331–355.
Krobová, Tereza & Jan Zàpotocký. 2022. I am not racist, but . . .: Rhetorical fallacies in arguments about the refugee crisis on Czech Facebook. *Journal of Intercultural Communication* 21(2). 58–69.
Langton, Rae. 2018. The authority of hate speech. In John Gardner, Leslie Green & Brian Leiter (eds.), In *Oxford Studies in Philosophy of Law*, 3rd ed., 123–152. Oxford: Oxford University Press.
Langton, Rae. 2012. *Beyond belief: Pragmatics in hate speech and pornography*. Cambridge: Cambridge University Press.
Ljubešić, Nikola, Darja Fišer & Tomaž Erjavec. 2019. The FRENK datasets of socially unacceptable discourse in Slovene and English. In *Text, Speech, and Dialogue: 22nd International Conference, Ljubljana, Slovenia, September 11–13, 2019, Proceedings 22, 103–114*. New York: Springer.
Magu, Rijul & Luo Jiebo. 2018. Determining code words in euphemistic hate speech using word embedding networks. In *Proceedings of the 2nd Workshop on Abusive Language Online (ALW2)*, 93–100. Association for Computational Linguistics.
Matsuda, Mari. 1989. Public response to racist speech: Considering the victim's story. *Michigan Law Review* 87(8). 2320–2381.
Matsuda, Mari, Charles R. Lawrence, Richard Delgado & Kimberlé Crenshaw. 1993. *Words that wound: Critical race theory, assaultive speech, and the first amendment*. Boulder: Westview Press.
Maitra, Ishani & Mary-Kate McGowan. 2007. The limits of free speech: Pornography and the question of coverage. *Legal Theory* 13(1). 41–68.
Musolff, Andreas. 2007. Which role do metaphors play in racial prejudice? The function of anti-Semitic imagery in Hitler's Mein Kampf. *Patterns of Prejudice* 41(1). 21–44.
Musolff, Andreas. 2015. Dehumanizing metaphors in UK immigrant debates in press and online media. *Journal of Language Aggression and Conflict* 3(1). 41–56.
Nicolas, Loïc. 2014. L'évidence du complot: Un défi à l'argumentation Douter de tout pour ne plus douter du tout. *Argumentation et analyse du discours* 13. 1–14.
O'Brien, Gerald V. 2003. Indigestible food, conquering hordes, and waste materials: Metaphors of immigrants and the early immigration restriction debate in the United States. *Metaphor and Symbol* 18(1). 22–47.
Oz Mustafa, Pei Zheng & Gian Chen. 2018. Twitter versus Facebook: Comparing incivility, impoliteness, and deliberative attributes. *New Media & Society* 20(9). 3400–3419.
Plenta, Peter. 2020. Conspiracy theories as a political instrument: Utilisation of anti-Soros narratives in Central Europe. *Contemporary Politics* 26(5). 512–530.
Popper, Karl. [1945] 1994. *The open society and its enemies*. Princeton: Princeton University Press.
Santa Ana, Otto, 1999. "Like an animal I was treated": Anti-immigrant metaphor in US public discourse. *Discourse & Society* 10(2). 191–224.
Sanguinetti, Manuela, Fabio Poletto, Cristina Bosco, Viviana Patti & Marco Stranisci. 2018. An Italian Twitter corpus of hate speech against immigrants. In *Proceedings of the Eleventh International Conference On Language Resources And Evaluation*(LREC 2018), 2798–2895. European Language Resources Association.
Semino, Elena, Zsófia Demjén & Jane Demmen. 2018. An integrated approach to metaphor and framing in cognition, discourse, and practice, with an application to metaphors for cancer. *Applied Linguistics* 39(5). 625–645.

Stephan, Walter, Oscar G. Ybarra & Kimberly Morrison Rios. 2016. Intergroup threat theory. In Todd D. Nelson (ed.), *Handbook of prejudice, stereotyping, and discrimination*, 255–278. London: Psychology Press.

Taguieff, Pierre-André. 2005. *La foire aux 'Illuminés': Ésotérisme, théorie du complot, extrémisme*. Paris: Mille et une nuits.

Waldron, Jeremy. 2012. *The harm in hate speech*. Cambridge: Harvard University Press.

Warner, William & Julia Hirschberg. 2012. Detecting hate speech on the World Wide Web. In *Proceedings of the Second Workshop on Language in Social Media, Montréal, Canada, June 7 2012*, 19–26. Montreal: Association for Computational Linguistics.

Wohlfeld, Monika. 2014. Is migration a security issue? In Omar Grech & Monika Wohlfeld (eds.), *Migration in the Mediterranean: Human rights, security and development perspectives*, 61–77. Msida: Mediterranean Academy of Diplomatic Studies.

Wodak, Ruth. 2015. *The Politics of fear: What right-wing populist discourses mean*. London: Sage publications.

Victoria Guillén-Nieto
Chapter 8
The wording of hate speech prohibition: "You can't see the wood for the trees"

1 Introduction

Migration tends to be represented as a crisis without a face. "The umbrella term *migrant* [. . .] has evolved from its dictionary definitions into a tool that dehumanises and distances, a blunt pejorative" (Malone 2015 in Monnier, Boursier & Seoane 2022b: 7). Migrants are commonly an object of hate speech, often encoded in the racial insult. The racial insult, according to Delgado, "remains one of the most pervasive channels through which discriminatory attitudes are imparted. Such language injures the dignity and self-regard of the person to whom it is addressed, communicating the message that distinctions of face are distinctions of merit, dignity, status, and personhood" (Delgado [1993] 2018: 90). Hate speech against migrants becomes a serious barrier for social integration and respect for diversity in modern societies (cf. Arcila-Calderón et al. 2022).

This chapter focuses on the legal approaches to hate speech prohibition designed to protect vulnerable groups such as migrants. From a linguistic perspective, I address the major problems raised by the wording of hate speech prohibition at three levels: international law, common law and civil law, especially European law and Member State law. The study aims to answer three questions: (1) Why is it difficult to interpret the current legal approaches to hate speech prohibition? (2) Can linguistic theory improve our understanding of the law's wording concerning hate speech prohibitions? (3) If so, which theory from the linguistic toolbox may be the most effectual to undertake the venture?

First, the present chapter begins by explaining how migration and hate speech are grounded. Second, a selected sample of legal approaches to hate speech prohibition is analysed to identify the major problems posed to legal meaning interpretation by the law's wording. Third, the analysis moves on to describe and explain hate speech from a linguistic perspective. This action involves two operations: (1) identifying the speech acts typically employed in the technical legal prohibitions of hate speech and (2) giving an account of the pragmatic cohesion of such speech acts at a macro level. In other words, I will explain how the speech acts associated with hate speech are organised linearly in a sequence and hierarchically in a global speech

Victoria Guillén-Nieto, University of Alicante

act. The chapter also looks at the problem of hate speech as a *complex speech act* at a micro level.

2 Migration and hate speech

Poverty, natural disasters, political persecution and war make people face danger, experience fear and react with what psychologists call a "fight-or-flight" response. In other words, whereas some people may decide to stay and deal with adversity and fight off the enemy, others would rather emigrate to other countries to start a new life. As noted in Chapter 1 in this volume, it has been empirically demonstrated that the emotion of fear is physiologically associated with the flight instinct, as demonstrated by Cannon (1915 [1925]: 187–188).

Migration to other countries is never easy; many obstacles can get in the way, such as family break-ups, loss of a home and all possessions, and various perils that may lead to death. Moreover, migrants have to negotiate formidable physical, administrative and legal barriers (see Chapter 3 in this volume). For example, let us take the case of Sub-Saharan Africans or Syrians embarking blindly on *pateras* ('open boats') or makeshift barges hoping to arrive on Southern European Mediterranean coasts to start new lives, or the case of Sub-Saharan migrants jumping over the high metal fence on the Moroccan-Spanish border to find better living conditions in Europe. Along the way, migrants become vulnerable; their despair and helplessness make them easy prey for mafia groups and human traffickers. Even when they evade such hazards, migrants can still be arrested or confined to ghettos while awaiting an administrative or judicial decision or can be repatriated. Figure 8.1 shows the despair of Syrian migrants arrested in Greece in 2010.

Furthermore, when migrants arrive in other countries, they are often perceived by locals as "a threat of both moral and physical nature" (Kopytowska, Woźniak & Grabowski 2017: 32). It is paradoxical that people who leave their home countries because they feel endangered become a "social threat" in the countries to which they flee.[1] The sense of a threat posed to long-established socio-economic status and political power can awaken or exacerbate protectionist patriotism, as well as the negative emotions of fear, resentment and anger[2] towards migrants. According to Kopytowska,

---

[1] For a consideration of the concepts of "danger" and "social threat", see Chapter 2 in this volume.
[2] Connan ([1915] 1925: 187–188) demonstrated that anger is associated with the instinct for fighting or attack. Anger embraces other related emotions such as hatred, rage, hostility, agitation, frustration and antagonism.

Chapter 8 The wording of hate speech prohibition — 175

Figure 8.1: Arrested refugees-immigrants in Fylakio detention centre, Evros, Greece, on 9 October 2010. Photo by Ggia.³

Woźniak and Grabowski, "Patriotism is conceptualised and used as motivation for and justification of hate speech directed against refugees and migrants" (2017: 34).

Apart from this association with negative emotions, hate speech is a surface manifestation of cognitive processes, social practices and of consciously discriminatory discursive choices.⁴ With regard to the former, social prejudice, intolerance and negative stereotypes (Ivanou 2017: 29–32) work to deny their targets access to mainstream society. These negative representations and shared meanings can even lead the vulnerable groups "to believe [the] frequent accusations that they are lazy, ignorant, dirty, and superstitious" (Delgado & Stefancic 2018: 8). The social prejudices, intolerance and negative stereotypes underlying hate speech against migrants in Europe have been the object of recent studies by Ascone (2022: 63–81), Paasch-Colberg, Trebbe, Strippel and Emmer (2022: 137–163), Schröter (2022: 83–114) and Strani and Szczepaniak-Kozak (2022: 21–61).

Moreover, migrants are subject to social practices of discrimination, such as dehumanisation (Kopytowska, Woźniak & Grabowski 2017: 35), demonisation,

---

3 File: 20101009 Arrested refugees immigrants in Fylakio detention centre Thrace Evros Greece. tif Creative Commons License 3.0 Share-Alike.
4 For an analysis of discursive practices of discrimination, see Chapters 4 and 5 in this volume.

marginalisation and stigmatisation, motivated by bias against race, colour, nationality, religion and sexual orientation. Hate-advocating speakers also resort to specific discursive choices with a view to constructing ingroup cohesion and outgroup exclusion. Such discursive choices include, as explained by Baider, Constantinou and Petrou (2017: 38), epistemic modality, intensification and metaphors, among others. Specifically, on examining the metaphors related to othering refugees and migrants in Europe, Baider, Constantinou and Petrou (2017: 38–42) found that whereas animal metaphors such as *worms* and *mice* are used to dehumanise refugees, migrants tend to be demonised, being instead associated with *outlaw, violence* and *disease*. However, "Racism and racial stigmatisation", as Delgado and Stefancic claim, "harm not only victims and perpetrators but society as a whole. They contravene the ideal of egalitarianism, that 'all men are created equal'" (Delgado & Stefancic 2018: 10).

3 The wording of hate speech prohibition

Laws regulating the prohibition of hate speech against vulnerable groups are designed to keep this harmful type of discourse from escalating into something more dangerous, such as violent hate crimes. As Monnier, Boursier and Seoane argue, "hate speech is not illegal because it is hateful, but because it is dangerous: either because it opens on to discrimination and violence, or because it leads indirectly to it" (Monnier et al. 2022b: 5). Hate speech prohibition is therefore an abstract endangerment statute, because it punishes hypothetical social risk.

The legislation of hate speech prohibition is analysed in depth in the *Report of the United Nations High Commissioner for Human Rights on the Expert Workshops on the Prohibition of Incitement to National, Racial or Religious Hatred* (2013) (henceforth, *Report of the United Nations*) whose purpose is

> to provide a comprehensive assessment of the implementation of legislation, jurisprudence and policies regarding advocacy of national, racial or religious hatred that constitutes incitement to discrimination, hostility and violence at the national and regional levels while encouraging full respect for the fundamental right of freedom of opinion and expression as protected by international human rights law. (*Report of the United Nations* 2013: 1)[5]

The *Report of the United Nations* points to some problematic aspects relating to hate speech legislation, such as: (1) the lack of a unified technical legal definition of hate speech; (2) the heterogeneity of the current legal approaches to hate

---

5 https://www.ohchr.org (accessed 5 November 2022).

speech, ranging from excessively narrow to overly broad; (3) the uneven and *ad hoc* applications of hate speech laws; (4) the scarce jurisprudence on hate speech; (5) the difficult task of adequately balancing the right of freedom of expression with the prohibition of incitement to discrimination, hatred, hostility or violence; (6) the problems relating to curbing freedom of information and the use of the Internet; (7) the difficulty of determining which speech acts can be defined as incitement to discrimination, hatred, hostility or violence and, therefore, can be prohibited by law.

Technical legal definitions must be clearly and narrowly defined for lawyers to have a reasonable understanding of the meaning of legal concepts and ultimately assess potential liability in specific cases. As mentioned above, the need for a unified technical definition of hate speech is a major legal problem for the court in evaluating potential social risk in actual cases. This section analyses several legal approaches to hate speech prohibition at three levels: (a) International law, (b) Common law and (c) Civil law—European law and Member State law.

a) International law

At the international level, hate speech was first prohibited by law in the 1960s, when the civil rights movement was at its peak. Specifically, the International Convention on the Elimination of All Forms of Racial Discrimination was adopted by the United Nations General Assembly Resolution 2106 in 1965, and the International Covenant on Civil and Political Rights was adopted by the United Nations General Assembly Resolution 2200A (XXI) in 1966. In what follows, I will refer to these two international legal standards in further detail.

The International Convention on the Elimination of All Forms of Racial Discrimination (1965) prohibits:

> All dissemination of ideas based on racial superiority or hatred, incitement to racial discrimination, as well as all acts of violence or incitement to such acts against any race or group of persons of another colour or ethnic origin, and also the provision of any assistance to racist activities, including the financing thereof. (Article 4, Section a)[6]

The above lists the types of hateful conduct that are sanctionable by law. Arguably, this legal approach to hate speech is an integrative one—it is content-based (dissemination of white supremacism), intent-based (incitement to hatred, discrimination or violence) and harms-based (acts of violence against the target groups). The limitation of freedom of expression is narrowly defined, and is consistent with Article 19 of the Universal Declaration of Human Rights[7] because the

---

6 https://www.ohchr.org (accessed 5 November 2022).
7 https://www.ohchr.org (accessed 5 November 2022).

restrictions imposed are meant to protect individuals and communities against a legitimate interest: racial hatred. Although the law's wording refers to disparate wrongful acts, it stresses "incitement to racial discrimination or acts of violence" as the main acts of hate speech. However, that the law's wording fails to make a clear distinction between a hate speech crime and a hate crime can be seen in this statement: "as well as all acts of violence or incitement to such acts against any race or group of persons of another colour or ethnic origin". As a result, the purpose of regulating hate speech, which is to prevent it from escalating into hate crime, is blurred.

On the other hand, the International Covenant on Civil and Political Rights (1966), in Article 20, paragraph 2, prohibits: "Any advocacy of national, racial or religious hatred that constitutes incitement to discrimination, hostility or violence".[8] This prohibition is also clearly and narrowly defined, consistent with Article 19 of the Universal Declaration of Human Rights, because it serves a legitimate interest: to protect the civil rights of individuals and communities against national, racial or religious hatred. The law's wording is intent-based, that is, based on the speaker's intentionality. Interestingly, the prohibition is expressed through a restricted modifying relative clause: "Any advocacy of national, racial or religious hatred *that constitutes incitement to discrimination, hostility or violence*" (my emphasis). This relative clause restricts the sentence meaning so that the law only prohibits "Any advocacy of national, racial or religious hatred" if and when it "constitutes incitement to discrimination, hostility or violence". The law's wording, however, does not resolve the question of how one wrongful act—"advocacy of national, racial or religious hatred"—can be interpreted as another wrongful act—"incitement to discrimination, hostility or violence". In section 4.2, I will try to throw light on this legal problem using the linguistic concept of *complex speech act* (cf. Guillén-Nieto 2023: 123).

Although adopted one year later than the International Convention on the Elimination of All Forms of Racial Discrimination (1965), the wording of the International Covenant on Civil and Political Rights (1966) offers some conceptual improvements. First, it clearly prohibits hate speech (abstract endangerment) instead of hate crime (actual harm). Second, the actions that the hate-advocating speaker can incite are presented in a three-grade sequence: discrimination, hostility or violence. This logical gradation depicts the process through which hate speech may escalate into acts of violence depending on the context of situation, which comprises, among other elements: the socio-political context, the participants involved, the medium, the type of statement, and the purpose. Third, the law's wording extends its protective mantle to individuals and communities against national, racial and religious

---

8 https://www.ohchr.org (accessed 5 November 2022).

hatred. Hence, the legal protected characteristics are enlarged compared to those listed in the International Convention for the Elimination of All Forms of Racial Discrimination (1965).

b) Common law

In the United States hate speech is not explicitly regulated by law and the Constitution requires a high threshold for restricting freedom of expression. Specifically, an individual's speech cannot be restricted under the law unless it meets the three criteria of the so-called *imminence test:*[9] (1) the speech is intended to produce lawless action; (2) the speech is likely to incite such action; (3) such action will likely incite an immediate breach of the peace. The imminence test foregrounds the speaker's malicious intentionality (*illocutionary act*) and the likely and immediate effects it may have upon the targets (*perlocutionary act*). In actual cases of hate speech, especially when delivered online, the above three criteria can be very difficult, or even impossible, for the court to elucidate. More recently, online hate speech has opened a legal debate about the suitability of the imminence test as a legal standard for evaluating cases associated with hate speech. According to Brenner, the application of the imminence test "makes it more difficult for a jury to convict the sender of an online hate message or remark because the connection between words broadcast over the Internet and the reader's reaction is difficult to gauge" (Brenner 2002: 781). After the speaker has made hateful statements over the Internet, it is difficult to say when readers will receive them, who will actually read them and what the effects upon vulnerable groups will be. The audience's reactions to hate speech are always unpredictable. As an alternative, Delgado (1982) proposed a *no-imminence* standard according to which hate-advocating Internet users should bear some civil liability if a causal link can be found between their advocacy of hatred and subsequent harm inflicted on members of the target group. More specifically, Delgado proposed regulating online hate speech through the torts of intentional infliction of emotional distress and group defamation. In his view, tort law can supply models for harm-based codes compatible with existing restrictions on the right to freedom of opinion and expression under the First Amendment of the United States Constitution.

Although in other common law jurisdictions, such as Canada, Australia and the United Kingdom, there is no specific hate speech law, the criminal code explicitly bans incitement to hatred. Specifically, the Canadian Criminal Code[10] prohibits

---

9 This test is also known as the *Brandenburg test* because it resulted from *Brandenburg v. Ohio* (1969), considered a landmark case in US case law.
10 Criminal Code, R. S. C. 1985, c. C-46, s. 319(1). https://laws-lois.justice.gc.ca (accessed 5 November 2022).

"incitement to hatred" and "wilful promotion of hatred" against any group identifiable by legally protected characteristics, such as race, nationality, sexual orientation or religion, in these terms:

(1) Everyone who, by communicating statements in any public place, incites hatred against any identifiable group where such incitement is likely to lead to a breach of the peace is guilty of (a) an indictable offence and is liable to imprisonment not exceeding two years, or (b) an offence punishable on summary conviction.
(2) Everyone who, by communicating statements other than in private conversation, wilfully promotes hatred against any identifiable group is guilty of (a) an indictable offence and is liable to imprisonment for a term not exceeding two years; or (b) an offence punishable on summary conviction. (Canadian Criminal Code, Section 319)

The Canadian Criminal Code provides a clear and narrow definition consistent with Article 19 of the Universal Declaration of Human Rights (1948) and Article 19 of the International Covenant on Civil and Political Rights (1966) because it only restricts speech that constitutes incitement to hatred where such incitement is likely to lead to violent acts, and wilful promotion of hatred when communicating the statements in public. Hence, it can be said that the Canadian Criminal Code prohibits an individual's speech if the speaker uses a public medium with the intent to communicate hateful messages (2), and the perlocutionary effect of such communication is damage or violence as ways of a breach of the peace (1). Nevertheless, the Canadian legal approach fails to explain how the two wrongful acts that fall under the category of hate speech—"incitement to hatred" and "wilful promotion of hatred"—are interlinked. Besides, the law's wording is overly broad regarding how "wilful promotion of hatred" and "incitement to hatred" can be perpetrated ("communicating statements"), about the locus ("any public place"), the target groups ("any identifiable group"), and even about the ways the peace can be broken ("such incitement is likely to lead to a breach of the peace". The abuse of general statements (*generalisation*) and the indefinite determiner *any* in the law's wording detract from its precision.

The Public Order Act 1986 prohibits racial hate speech in the United Kingdom.[11] In Part III, Section 17, the law's wording explicitly defines racial hatred in

---

[11] Religious hatred joined the Public Order Act in 2006, and hatred on sexual orientation grounds in 2008. These adjustments demonstrate how social change may affect the law's wording and how the law's wording itself can contribute to changing public behaviour towards or against vulnerable groups, and even promote social change in several ways, including the use of sanctions or, more subtly, by channelling public behaviour through default rules.

these terms: "racial hatred means hatred against a group of persons defined by reference to colour, race, nationality (including citizenship) or ethnic or national origin".[12] Section 18 sets out the wrongful acts intended or likely to stir up racial hatred:

> (1) A person who uses threatening, abusive or insulting words or behaviour or displays any written material which is threatening, abusive or insulting is guilty of an offence if (a) he intends thereby to stir up racial hatred or (b) having regard to all the circumstances, racial hatred is likely to be stirred up thereby.

The Public Order Act then points to the speaker's communicative intent (illocutionary act)—the speaker "intends to stir up racial hatred"—and the perlocutionary effect—"racial hatred is likely to be stirred up". Besides, the law's wording indicates the wrongful acts that the speaker can perform to stir up hatred—"threatening, abusive or insulting words or behaviour or displays any written material [. . .]". It is arguable that these wrongful acts are auxiliary to the main act: "incitement to racial hatred". However, the law's wording is overly broad regarding the forms of expression that constitute hate speech and should be prohibited by law. Note that the law's wording is, again, built on generalisation and uses the indefinite determiner *any*. It is unclear which forms of expression the speaker can use to threaten, abuse or insult to incite hatred towards or against the targets. Undoubtedly, making an inventory of threatening, abusive or insulting forms of expression that may incite hatred could help pave the way for hate speech prohibition (see Chapter 6 in this volume). Section 4.2 of this chapter, however, explains the intrinsic difficulties that may thwart such an endeavour.

In Australian legislation, Section 77 of the Criminal Code Amendment Act (Racial Vilification Act 2004) prohibits conduct intended to incite "racial animosity" or "racial harassment", in these terms:

> Any person who engages in any conduct, otherwise than in private, by which the person intends to create, promote or increase animosity towards, or harassment of, a racial group or a person as a member of a racial group is guilty of a crime and is liable to imprisonment for 14 years.[13]

Interestingly, the wording of the Racial Vilification Act does not include the terminology recommended by international legal standards—"incitement to discrimination, hatred, hostility or violence", as it uses new key terms—"create, promote

---

12 The Public Order Act 1986. https://www.legislation.gov.uk (accessed 5 November 2022).
13 Criminal Code Amendment Act (Racial Vilification Act 2004). Section 77. https://www.legislation.wa.gov.au (accessed 5 November 2022).

or increase racial animosity towards, or harassment of, a racial group". The wording is overly broad about the type of conduct that is likely to create, promote or increase racial animosity or harassment. Note the use of the indefinite determiner *any* in "any conduct, otherwise than in private, by which the person intends to create, promote or increase animosity towards, or harassment of, a racial group". It should also be noted that the use of the verb "increase" with racial animosity or harassment pragmatically presupposes that this type of negative behaviour is an "assumed common ground" (Brown & Yule 1983: 28) in Australian society. Besides, the law's wording only protects the targets against racial hatred, setting aside other widely recognised protected characteristics such as religion, sex, sexual orientation and disability.

c) Civil law—European law

The ECRI's General Policy Recommendation No. 15 on Combating Hate Speech (henceforth, ECRI's Recommendation) is considered the European legal standard. It sets out that hate speech is to be understood as

> the advocacy, promotion or incitement, in any form, of the denigration, hatred or vilification of a person or group of persons, as well as any harassment, insult, negative stereotyping, stigmatisation or threat in respect of such a person or group of persons and the justification of all the primary types of expression, on the ground of race, colour, descent, national or ethnic origin, age, disability, language, religion or belief, sex, gender, gender identity, sexual orientation and other personal characteristics or status.[14]

Predictably, the law's wording presents interpretative difficulties because of its lack of cohesion. The law's wording provides an exhaustive enumeration of elements, which do not cohere as a cogent articulation. Although each element belongs to the "language of discrimination" (Stollznow 2017), they do not form a unified whole because they refer to disparate categories. In particular, there is a cognitive element of discrimination (negative stereotyping), a social practice (stigmatisation), and there are speech acts (harassment, insult and threat). These disparate categories are connected with the disjunctive conjunction *or* as if they were possibilities or alternatives within the same category. In addition, the long list of legally protected characteristics finishes with a final generalisation ("and other personal characteristics or status") that would be difficult for the court to discern in actual cases dealing with hate speech.

The ECRI's Recommendation provides guidelines to the Council of Europe member states on how to counter hate speech. Member states must then accommodate

---

[14] Council of Europe. European Commission Against Racism and Intolerance (ECRI) 2015. https://www.coe.int (accessed 5 November 2022).

the guidelines to their own specific legal cultures. In the following, I provide two examples of the results of such negotiation. The former comes from Germany, the latter from Spain.

d) Civil law—Member State law

Although there is no specific hate speech law in Germany, Section 130, Incitement of masses,[15] of the German Criminal Code explicitly prohibits:

> (1) Whoever, in a manner suited to causing a disturbance of the public peace,
> 1. incites hatred against a national, racial, religious group or a group defined by their ethnic origin, against sections of the population or individuals on account of their belonging to one of the groups mentioned above or sections of the population, or calls for violent or arbitrary measures against them or
> 2. violates the human dignity of others by insulting, maliciously maligning or defaming one of the groups above, sections of the population or individuals on account of their belonging to one of the groups as mentioned above or sections of the population incurs a penalty of imprisonment for a term of between three months and five years.[16]

Evidently, the German Criminal Code only restricts an individual's freedom of expression if there is a danger of "causing a disturbance of the public peace" (perlocutionary effect). Although the law's wording specifies the wrongful acts leading to such a global effect, their pragmatic cohesion at a macro level requires explanation. Since incitement is not an *explicit performative*[17] but instead an *implicit performative*, the hate-advocating speaker must communicate intentionality (illocutionary act) indirectly through other speech acts such as "insulting", "maligning" and "defaming" (section 4.2 of this chapter explains the difference between the notions of *primary illocutionary act* and *secondary illocutionary act* in a *complex speech act*). However, it should be noted that these speech can at the same time be the perlocutionary acts resulting from incitement to hatred that the hate-advocating speaker's followers may perform against the object (target-group members). Third, "violates the human dignity of others" is the global perlocutionary effect, resulting from "insulting", "maligning" and "defaming", on the members of the target group. Insults and defamation are crimes typified under Division 14 Insult[18]

---

[15] Division 7 Offences against public order, Section 130 Incitement to masses. Strafgesetzbuch – StGB. Translation provided by Prof. Dr Michael Bohlander. The translation is completely revised and regularly updated by Ute Reusch. https://www.gesetze-im-internet.de/englisch_stgb/ (accessed 22 November 2022).
[16] https://www.gesetze-im-internet.de/englisch_stgb (accessed 22 November 2022).
[17] Because incitement to hatred is considered a wrongful act, the hate-advocating speaker cannot state explicitly and directly: "I incite you to hate xy".
[18] German Criminal Code, Division 14 Insult: Section 185 Insult and Section 187 Defamation. https://www.gesetze-im-internet.de/englisch_stgb (accessed 22 November 2022).

of the Criminal Code. The circumstance that the victims of such offences are members of a legally protected group is considered an aggravating factor. In other words, such a circumstance raises the severity and punishment for the aggravated offence. Apart from incitement to racial and religious hatred, Section 130 also prohibits disseminating hateful content, Holocaust denial and Nazi Party approval or glorification.

The second example comes from Spanish civil law. As in Germany, there is no specific hate speech law in Spain. However, Article 510 of the 2015 Spanish Criminal Code (Section 1, paragraph a) sets the prohibition of wrongful acts that can fall under the category of hate speech.

> Those who, directly or indirectly, foster, promote or incite hatred, hostility, discrimination or violence against a group, or part thereof, or against a certain person for belonging to such a group, for reasons of racism, antisemitism or for other reasons related to ideology, religion or beliefs, family circumstances, the fact that the members belong to ethnicity, race or nation, national origin, gender, sexual orientation or identity, or due to gender, illness or disability (Criminal Code 2016: 214).[19]

Although the influence of ECRI's Recommendation on the prohibition of incitement to hatred in the Spanish Criminal Code is apparent here, it is interesting to note that Article 510 broadens the list of legally protected categories that the former provides: "racism, antisemitism or for other reasons related to ideology, religion or beliefs, family circumstances, the fact that the members belong to an ethnicity, race or nation, national origin, gender, sexual orientation or identity, or due to gender, illness or disability" (Criminal Code 2016: 214). The broadness of Article 510 also affects the category or element of "dissemination" (Section 1, paragraph b) to the extent that it incorporates more limitations to the right of freedom of opinion and expression:

> Those who produce, prepare, possess to distribute, provide third parties access to, distribute, publish or sell documents or any other type of material or medium that, due to the content thereof, are liable to directly or indirectly foster, promote or incite hatred, hostility, discrimination or violence against a group (Criminal Code 2016: 214).

In a similar vein to Section 130 of the German Criminal Code, Article 510 (Section 1, paragraph c) also refers to other wrongful acts that the legal concept of hate speech may encompass.

---

[19] All quotes from Article 510 are borrowed from the official English translation, Criminal Code 2016, edited by the Ministry of Justice and Technical Secretariat and translated by Clinter Traducciones e Interpretaciones. https://www.mjusticia.gob.es (accessed 22 November 2022).

Those who publicly deny, seriously trivialise or extol the crimes of genocide, crimes against humanity or against persons and property protected in the event of armed conflict, or who extol the perpetrators thereof, if committed against a group or part thereof, or against a certain person for belonging to such a group, for reasons of racism, antisemitism or for other reasons related to ideology, religion or beliefs [. . .] if such conduct promotes or encourages a climate of violence, hostility, hatred or discrimination against such individuals (Criminal Code 2016: 214).

In Section 2, paragraph a), Article 510 prohibits damage to human dignity (perlocutionary act) by any means of public expression and dissemination (public medium). Paragraph b) of the same section penalises exaltation or justification of offences (illocutionary act) committed by another group against the target groups by any means of public expression or dissemination (public medium).

Many jurists in Spain seem unhappy with the wording of Article 510 (see Landa-Gorostiza 2018; Doval Pais in Chapter 9 in this volume) because it can be difficult for the court to apply in actual cases of hate speech. Furthermore, its broadness means that it can invite differing rulings in the same or similar cases of hate speech.

In conclusion, the current legal approaches to hate speech prohibition are difficult to interpret, partly because the law's wording includes generalisations, indefinite determiners, disjunctive conjunctions and long enumerations of disparate categories and speech acts—and partly because it fails to explain the pragmatic cohesion between these disparate categories and speech acts. The general impression is that "you can't see the wood for the trees". That is to say, one may be too involved in the particulars of hate speech—of individual speech acts—to apprehend the gist of hate speech as a global speech act.

4 Hate speech through the lens of pragmatics

In his seminal work *How to do things with words* (1962), Austin laid the pragmatic foundation of Speech Act Theory, which explains how words can be used not only to report states of affairs but also to perform actions. Later, Brown and Yule clarified the value of Speech Act Theory in discourse analysis in these terms: "it provides an account of how some formally unconnected utterances go together in conversational discourse to form a coherent sequence" (Brown & Yule 1983: 233). In the same line of thought, van Dijk (1992) argued that describing the pragmatic structure of discourse means giving an account of its cohesion, that is, how speech acts are organised linearly in sequences and hierarchically in global speech acts. A discourse is coherent when the speech acts hang together to accomplish one global speech act. Meizhen and Yadi (2017: 57–82) elaborated on the concept of global speech act in their pragmatic analysis of courtroom discourse. In what follows, I will explain how the disparate categories and speech acts that fall within the wording of hate speech prohibition hang together to accomplish a super goal.

4.1 The superstructure and macrostructure of hate speech

As I claim in an earlier publication (Guillén-Nieto 2023: 109–111), the wrongful acts that fall within hate speech are pragmatically interconnected. Such acts can be explained as components of a sequence of speech acts that is organised both linearly and hierarchically. The sequence makes up the superstructure (scaffolding) of hate speech that sustains its cohesion—its unity of form—at a macro level.

Arguably, the components of the linear sequence of speech acts configuring hate speech can be classified into categories that I call *component major acts*: (a) hatred dissemination; (b) hatred advocacy; (c) incitement to hatred; (d) group harassment. These component major acts are also hierarchically organised, as each one performs a specific action contributing to the realisation of other component major acts in the sequence. The component major acts can together be assigned one global intention or plan (macrostructure), giving coherence or unity of meaning to the sequence. With hate speech, that global intention or goal is: to engage the audience in unlawful acts against members of the target group and, therefore, secure their exclusion from mainstream society or, in extreme cases, provoke their genocide. In the following, I will refer in greater detail to each of the component major acts of the sequence in turn.

4.1.1 Hatred dissemination

Propaganda, according to Tsesis, is "essential for eliciting widespread cultural acceptance of exclusionary and supremacist ideologies" (Tsesis 2002: 1). The speaker resorts to disseminating or spreading racist values, attitudes and thoughts about target-group members in a public medium, in order to potentially reach a large audience. In the hierarchy, hatred dissemination can be categorised as a preparatory act whose result is a necessary condition for the success of other major acts in the sequence. The other major acts can include: hatred advocacy, incitement to hatred, and group harassment. In other words, without dissemination the hate speech machine cannot be set in motion.

In turn, as a component major act, hatred dissemination also embraces other minor acts such as financing racist activities, provision of assistance to racist activities, advertising, promotion, and wilful promotion. These acts can be categorised as *component minor acts.* Together they can be assigned a common goal: hatred dissemination.

4.1.2 Hatred advocacy

The speaker will publicly advocate, support or defend racist values, attitudes and thoughts towards or against members of the target group. I concur with Tsesis that "Hatred directed against community members significantly influences peoples' perspectives, attitudes, and interpersonal relations. Preconceived notions about vulnerable minorities influence how they are treated" (Tsesis 2002: 9). Hatred advocacy can therefore be described as an auxiliary act whose successful result is a sufficient condition for the success of the next act in the sequence of component major acts: incitement to hatred.

Hatred advocacy encompasses other minor acts referred to in the wording of hate speech prohibition, such as: foster, support, justify, deny, glorify, approve, defend. These acts can be categorised as component minor acts. Together, they can be assigned a common goal: hatred advocacy.

4.1.3 Incitement to hatred

A crisis context can easily turn hatred advocacy into incitement to hatred or can be interpreted as incitement to hatred by any member of the targeted audience, even if this was not the speaker's communicative intention. I claim that incitement to hatred can be categorised as the main act of the sequence of component major acts. Let me give some substance to this claim. From the discussion in section 3 of this chapter, the reader will remember that the speech act incitement to hatred is present in all the sampled legal approaches to hate speech prohibition, with the sole exception of the Racial Vilification Act (2004) in Australia. Largely, the wording makes a focal point of incitement to hatred against vulnerable groups (cf. Chapter 9 in this volume). Interestingly, the verb *incite* is, according to Austin (1962), an *implicit performative.*[20]

---

[20] The concept of *implicit performative* originates in Austin's (1962) observation that while sentences can be used to report affairs (e.g. *The sun rises in the East and sets in the West*), the utterance of some sentences can be treated as the performance of an act (e.g. *I declare you husband and wife*). Originally, Austin named the former *constative speech acts* and the latter *performative speech acts*. Later, Austin concluded that the utterance of any sentence could be treated as the performance of an act. As a result, he reformulated the original classification into *explicit* and *implicit performatives.*

The hate-advocating speaker knows that it is socially reprehensible to say: "I incite you to perform unlawful acts against other people". Hence, for face-saving purposes, the call for unlawful action will not be performed explicitly but implicitly and, depending on the external circumstances, directly or—as is increasingly the case in online hate speech—indirectly (Vella Muskat & Assimakopoulos 2017: 42–46; Ruzaite 2018: 93–18; Becker 2020: 32–40, 2021; Baider 2022, Baider in Chapter 7 in this volume; Guillén-Nieto 2023: 119–129). Under Searle's classification (1969), incitement is a *directive speech act*—it is an utterance or statement that aims to get someone to do something. In the case of hate speech, the incitement is to commit a wrongful act.

Incitement encompasses other acts, often included in the wording of hate speech prohibition, such as encouraging and stirring up hatred. These acts can be categorised as *component minor acts*. Together, they can be assigned a common goal: incitement to hatred.

4.1.4 Group harassment

Austin (1962) pointed out that a speaker also performs a perlocutionary act when making an utterance. The perlocutionary act refers to the illocutionary act's *effect* on the hearer. I here claim that *group harassment* can be categorised as the perlocutionary act resulting from the main act—incitement to hatred (illocutionary act)—of the sequence of component major acts described in the preceding sections. Group harassment can then be described as a consequent act in the sequence of component major acts. However, the question is: why should group harassment, instead of, for example, insult, defamation or threat, also specified in the wording of hate speech prohibition, be the consequent act in the sequence of component major acts? In what follows, I will explain the reasoning behind my claim.

I will start by defining group harassment. The first scholarly definitions of harassment come from the field of social psychology. Leymann (1990) studied harassment in the workplace, which he named *mobbing*. He explained mobbing as a social phenomenon in which one or more perpetrators direct acts of hostile and unethical communication at a target—or targets—over a significant period, with the aim of nullifying their dignity and excluding them from the organisation. In an earlier publication of my own (Guillén-Nieto 2022) I defined harassment from a linguistic perspective, as a super-genre. I proposed five functional criteria for the classification of different sub-genres of harassment: (1) the locus; (2) the number of perpetrators; (3) the power relation between the perpetrator(s) and the target; (4) the identity of the target; (5) the mode of expression. According to these criteria, group harassment can be described as a sub-genre of harassment that: (1) takes place in the context of public life; (2) requires coordinated action by several individuals; (3) has a mixed nature because it can involve a combination of both vertical (from superior to inferior) or horizontal (between peers) types of harassment; (4) is discriminatory because the targets are identifiable groups of people, on the grounds of legally protected characteristics;[21] (5) can be expressed through any form of expression and public medium.

Unlike other consequent acts, such as insult, defamation and threat, group harassment is a macro act whose superstructure and macrostructure can be described. Group harassment encompasses other acts specified in the wording of hate speech prohibition, such as racial discrimination, increased racial animosity, hostility, insult, vilification, defamation, threat, denigration, acts of violence and violation of human dignity. The intensity, frequency and duration of these acts are used as measures to quantify the degree of hate speech's severity and evaluate potential social risk. These acts can be categorised as *component minor acts*. Together, they constitute group harassment and can be assigned a common goal, namely to secure the social exclusion of the targets.

One significant problem in arguing cases of group harassment before a court of law is the discontinuous structure of harassment. Group harassment requires the coordinated action of several or more individuals, through a succession of wrongful acts that may not be adjacent in time and space. Furthermore, group harassment may develop over an extended period. However, group harassment can still be described as a unified text that obeys a global strategic plan to pursue a global malicious purpose (cf. Guillén-Nieto 2022; Stein 2022).

The superstructure and macrostructure of hate speech as a global act with a super goal is diagrammed in Figure 8.2, showing the component major acts of the sequence: (1) Dissemination; (2) Advocacy; (3) Incitement; (4) Group harassment.

4.2 Hate speech at a micro level: a complex speech act

Brown and Yule (1983) pointed to several problems in applying Speech Act Theory in discourse analysis. The first problem concerns how to avoid arbitrariness in identifying speech acts in discourse. Brown and Yule argued that the problem of how to assign speech acts in a non-arbitrary way should "lead the analyst to recognise that the way speech acts are conventionally classified into discrete act-types such as 'request', 'promise', 'warn', etc. may lead to an inappropriate view of what speakers do with utterances" (Brown & Yule 1983: 233). One should then abandon the idea that a speech

---

[21] For an in-depth analysis of the origins of German anti-Semitism and racism in the United States, see Tsesis (2002: 9–65).

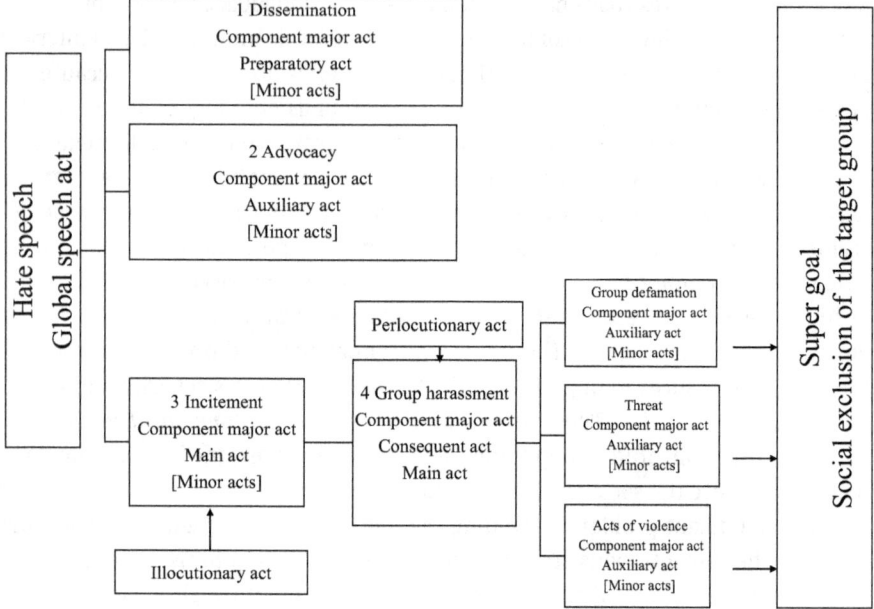

Figure 8.2: The superstructure and macrostructure of hate speech as a global speech act.

act is confined to a discrete utterance, as is the case with *I threatened you* or *I know where you live*. In the former, the directive speech act *threaten* is conveyed explicitly, while in the latter, it is conveyed implicitly. Based on Brown and Yule's reasoning, I conclude that from either the speaker or the hearer's point of view, several utterances strung together can constitute a single speech act. Example 8.1 below, which reproduces an excerpt from *Brandenburg v. Ohio* (1969), exemplifies this linguistic phenomenon.

Example 8.1
This is an organisers' meeting. We have had quite a few members here today which are—we have hundreds, hundreds of members throughout the State of Ohio. I can quote from a newspaper clipping from the Columbus, Ohio, Dispatch, five weeks ago Sunday morning. The Klan has more members in the State of Ohio than does any other organisation. We're not a revengent [sic] organisation, but if our President, our Congress, our Supreme Court, continues to suppress the white, Caucasian race, it's possible that there might have to be some revengeance [sic] taken. (*Brandenburg v. Ohio*, 395 US 4444 (1969), p. U.S. 446).

The whole excerpt implicitly conveys a *threat*. The wording of the threat is ambiguous, because it can be interpreted as a threat against African Americans and Jews (the target groups in the case), a threat against the state authorities for giving social rights to the targets, or a threat against the target groups and state authorities. The threat is implicitly conveyed through a mixed conditional: "if our President, our Congress, our Supreme Court, continues to suppress the white, Caucasian race [*condition*], it's possible that there might have to be some revengeance taken [*consequence*]. It should be noted that the consequence is strategically designed to hide the speech act's illocutionary force [*threat*]. The wording strategically includes the following linguistic elements to hide agency:

> [*impersonal structure*]: *It's + possible (modal adjective)* [*epistemic modality*][22] + *that clause* [*focus*] *there* [*impersonal structure*] *might (modal verb)* [*epistemic modality*] *have to be some revengeance taken (passive voice)* [*impersonal structure*]

The result was that Brandenburg's conviction was overturned by the Supreme Court of the United States. The grounds for the overturning were that the local government could not constitutionally punish abstract advocacy of force or law violation unless directed to incite or produce imminent lawless action, and such advocacy is likely to produce or incite such action. This court decision set the new legal standard for hate speech in US case law: the imminence test, also known as the Brandenburg test.

A second problem lies in the fact that one statement—or utterance—may perform several speech acts simultaneously. Speakers can say one thing and mean it while at the same time meaning another thing—or even several things at once. This linguistic phenomenon is known as a *complex speech act* (Beyssade & Marandin 2006). It can be described as the simultaneous performance of two acts: (1) what speakers commit themselves to and (2) what speakers call on the hearer to perform. The hearer's task is then to infer which is the *primary illocutionary act* (the intended act) and which is the *secondary illocutionary act*(s) (the act or acts expressed) in a given context.

The concept of a complex speech act can improve understanding of hate speech at a micro level. Specifically, in the case of hate speech the primary illocutionary speech act is a directive—incitement to hatred. The secondary illocutionary act can be of various types, e.g. representative, commissive, directive and expressive (Searle 1969, 1975, 1979).[23] An illustrative example of hate speech as a

---

22 Epistemic modality refers to the conveyance of the speaker's attitude towards the factualness of a proposition. Epistemic modality is linguistically encoded in modal verbs, adjectives, adverbs and other grammatical elements. In Example 8.1, Brandenburg exhibits his linguistic mastery. He intentionally uses epistemic modality to mask the threat. Brandenburg expresses his attitude about the factualness of the proposition, accepting that there is a possibility, but not being certain.
23 Representative speech acts commit the speaker to something being the case. Directive speech acts try to make the hearer do something. Commissive speech acts commit the speaker to do something in the future. Expressive speech acts express how the speaker feels about the situation.

complex speech act can be seen in Article 20, paragraph 2 of the International Covenant on Civil and Political Rights (1966). This article prohibits "Any advocacy of national, racial or religious hatred that constitutes incitement". This means that hate-advocating speakers may be sincerely *advocating* hatred at the same time that they are actively *inciting* hatred. Hence, the primary illocutionary act is incitement (*directive speech act*), and the secondary illocutionary act is advocacy (*representative speech act*). In other words, incitement is performed indirectly, through advocacy. Figure 8.3 depicts hate speech as a complex speech act.

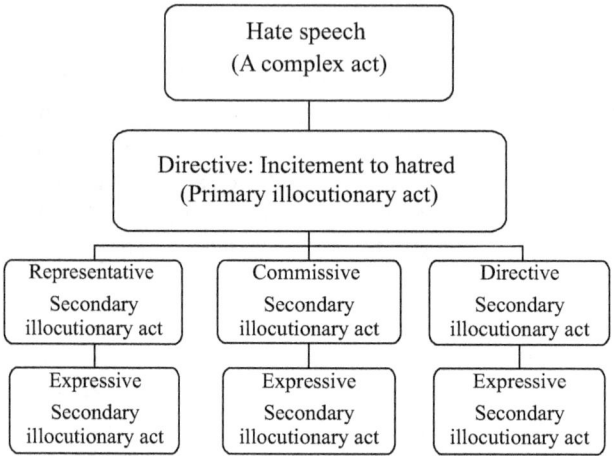

Figure 8.3: Hate speech as a complex speech act (Guillén-Nieto 2023: 123).

By way of illustration, let us consider Example 8.2, which reproduces abusive remarks about ethnic groups and immigrants made by three members of the Greenjackets in Østerbro (Copenhagen) in a broadcast interview on Danmarks Radio (Denmark). The remarks are borrowed from *Jersild v. Denmark* (1994), tried at the European Court of Human Rights.[24]

Example 8.2

[The] Northern States wanted that the niggers should be free human beings, man, they are not human beings, they are animals. [*Representative speech act*]

Just take a picture of a gorilla, man, and then look at a nigger, it's the same body structure and everything, man, flat forehead and all kinds of things. [*Representative speech act*]

---

[24] The European Court of Human Rights. Case of Jersild v. Denmark. Application No. 15890/89. Judgment. Strasbourg. 23 September 1994. Official English version of the judgment (p. 9).

A nigger is not a human being, it's an animal, that goes for all the other foreign workers as well, Turks, Yugoslavs and whatever they are called. [*Representative speech act*]

It is the fact that they are "Perkere", that's what we don't like, right, and we don't like their mentality [. . .] what we don't like is when they walk around in those Zimbabwe clothes and then speak this hula-hula language in the street. [*Expressive speech act*]

It's drugs they are selling, man, half of the prison population in "Vestre" are in there because of drugs [. . .] they are the people who are serving time for dealing drugs. [*Representative speech act*]

They are in there, all the "Perkere", because of drugs. [*Representative speech act*]

To begin with, Example 8.2 illustrates the pragmatic cohesion of the speech acts constitutive of hate speech:

1) Hatred dissemination: The journalist interviewing three members of the Greenjackets was charged with aiding and abetting the dissemination of the Greenjackets' racist views, and the same charge was brought against the head of the news section of Danmarks Radio.
2) Hatred advocacy: the Greenjackets publicly proclaimed racist views about immigrants.
3) Incitement to hatred: to prevent social risk or damage to the target group (group harassment), the Greenjackets were charged with incitement to racial hatred.

On the other hand, it is important to note that, although none of the abusive remarks included the directive speech act (incitement to hatred), the Public Prosecutor concluded that this was the primary illocutionary act (the intended act), instead of the representative and expressive speech acts performed by the speakers (secondary illocutionary acts). Example 8.2 therefore shows several utterances strung together which together constitute a single unified speech act.

The Greenjackets, Jersild and the head of the news section were convicted in the city court. Jersild and the head of the news section appealed against the city court's judgment to the High Court of Eastern Denmark. The appeal was dismissed. Subsequently, Jersild and the head of the news section appealed to the Supreme Court. Jersild claimed that the programme aimed to draw public attention to a new phenomenon in Denmark: racism and xenophobia as practised by socially disadvantaged youths. He explained that the purpose of including the Greenjackets' offensive statements in the programme was to counter them through public exposure. In his view, the offending remarks make a mockery of their authors, rather than bolstering their racist views.

The Supreme Court confirmed Jersild's conviction. In the justices' opinion, Jersild had failed to fulfil his duties and responsibilities as a journalist because he had irresponsibly encouraged the youths to make racist statements and had failed to appropriately counteract such statements in the programme.

Jersild lodged an application to the European Court of Human Rights (ECHR) claiming a violation of his right to freedom of expression under Article 10 of the Convention. This was the first time the ECHR had been concerned with a case of dissemination of racist remarks that deny the quality of human beings to a large group of people. At the heart of the deliberation was the question of how to balance Jersild's right to impart information versus the protection of the reputation of those who have to suffer racial hate speech. The majority opinion attributed more weight to Jersild's right to freedom of expression than to the rights of the targeted-group members, and hence Jersild's conviction was overturned by the ECHR.

An additional problem the analyst faces lies in the fact that speakers tend to resort increasingly to *implicitness* (Austin 1962)[25] and *indirectness* (Searle 1979: 30–57)[26] for the expression of hateful messages.[27] Using covert strategies, the hate-advocating speaker can avoid criminal prosecution and may be more successful in achieving their malicious communicative purposes. Incitement to hatred is commonly expressed through implicature (Grice 1975)—that is, the nefarious meaning is implied or suggested in a given context of situation. Although it could be legally useful to make an inventory of the forms of expression that can be used to advocate or incite hatred,[28] the task is bound to be unproductive because the meaning of hateful messages must be interpreted in the context of situation in which the utterance—or statement—was made.[29]

The context of situation provides the cues necessary to interpret the illocutionary force of an utterance, though sometimes there may be a mismatch between the speaker's communicative intention and the hearer's interpretation (Guillén-Nieto 2023: 124–128). Persons of similar ideological alignment as the hate-advocating speaker are more likely to correctly interpret incitement to hatred than others.

---

[25] The meaning is implied or understood though not directly expressed.
[26] The meaning is conveyed through another speech act other than that intended.
[27] For a consideration of covert hate speech, see Chapter 7 in this volume.
[28] For a taxonomy of hate speech forms, see Chapter 6 in this volume.
[29] The Rabat Plan of Action (2012) sets a high standard for limitations on freedom of expression, consisting of six criteria to determine where an expression creates such a danger of harm to justify prohibitions on expression. The proposed criteria are context-based: (1) the social and political context where the expression occurred, (2) the identity of the speaker—e.g. their status and influence over the audience, (3) the speaker's communicative intention, (4) the content and form of the expression, (5) the extent of the expression and (6) the likelihood and imminence of violence.

5 Conclusions

Technical legal definitions need to be clearly and narrowly defined, and must be sufficiently specific to enable lawyers to understand the meaning of legal concepts and thereby assess potential liability in specific cases. With respect to existing hate speech prohibitions, these show varying degrees of specificity, use diverse terms to refer to hate speech offences, and some arguably violate freedom of expression. The chapter showed that, in general, it is difficult to interpret the current legal approaches to hate speech, especially in European law and Member State law. The reason for this lies in the impreciseness of the law's wording. The texts of these laws include long enumerations of disparate categories, broad generalisations, and they overuse the indefinite determiner *any*.

I have tried to show how linguistic theory can improve understanding of the legal concept of hate speech. In particular, at a macro level, Speech Act Theory can give an account of the superstructure and macrostructure of hate speech as a global act. As a global act, hate speech pursues a super goal: the persecution, social exclusion and, in extreme cases, genocide of a vulnerable group. Specifically, I propose four component major acts that are organised linearly in a sequence and hierarchically in a global speech act: (1) hatred dissemination, (2) hatred advocacy, (3) incitement to hatred and (4) group harassment. This model can make legible the pragmatic cohesion of the major speech acts that constitute hate speech. Each major act comprises other minor acts whose micro goals contribute to a macro goal. All the macro goals together contribute to the super goal (the macrostructure of hate speech).

Speech Act Theory can also describe and explain hate speech at a micro level as a complex speech act. A complex speech act is double-faceted: (1) what speakers commit themselves to and (2) what speakers call on the hearer to perform. The hearer's task is to infer which is the primary illocutionary act (the intended act) and which is the secondary illocutionary act(s) (the act or acts expressed) in a given context. With hate speech, the primary illocutionary act (the intended act) is commonly performed implicitly and indirectly. In this way, the hate-advocating speaker tends to avoid social reprobation and is protected from criminal prosecution. Hence, there are inherent difficulties in correctly interpreting the speaker's intentionality or correctly evaluating the social risk in actual cases of hate speech. The expert linguist can provide technical assistance to the court concerning the interpretation of meaning, especially when this has been conveyed by way of implicature (contextual meaning). In addition, forensic linguistic analysis can offer substantial clues about the speaker's intentionality in a given context.

It is well-known that legally protected groups of people exist because they are burdened by a long history of discrimination, social exclusion, hostility and violence. The proposed explanation of hate speech as a global speech act in the pursuit of a super goal (persecution, social exclusion or, in extreme cases, genocide) also supports arguments for the civil liability of any subject contributing to the global act of hate speech against a minority group, in any form, in any place and at any particular period.

In sum, the present chapter invites reflection on the wording of hate speech prohibition (cf. Chapter 9 in this volume) and points to some grammatical problems that hinder expression and pose difficulties of legal interpretation. Specifically, the long enumerations of mixed categories, generalisation and indefinite determiners make the texts of hate speech laws overly vague and damagingly imprecise.

Through the application of Speech Act Theory, legal practitioners can come to understand how the confusingly disparate speech acts enlisted in the law's wording form a coherent pragmatic sequence of speech acts, extending over time, with cumulative effects on both the audience (the ingroups) and the object (the outgroups—the targets). With the sequence in mind, legal practitioners should be cautious about treating hate speech as an isolated act with no social repercussions. Members of a hate speech targeted group know that any manifestation is a link in a sequence, forming part of a global speech act that can have devastating effects on its victims. Even an isolated racial slur can conjure an entire history of supremacist ideology, racist discrimination, exclusion, hostility and violence.

As a last word, I will maintain that analysing the complex speech act that hate speech represents always requires thorough linguistic analysis, because its meaning is not utterance-based but context-based. Thanks to their special training, linguists can find linguistic traces pointing to the speaker's disguised intentionality that can be singularly useful to the court in making legal decisions.

References

Arcila-Calderón, Carlos, Patricia Sánchez-Holgado, Cristina Quintana-Moreno, Javier-J. Amores & David Blanco-Herrero. 2022. Discurso de odio y aceptación social hacia migrantes en Europa: Análisis de tuits con geolocalización. *Comunicar. Media Education Research Journal* 30(71). 21–34.

Ascone, Laura. 2022. Discrediting the other, building in-group bonds: An analysis of French and Italian user-generated hate contents. In Angeliki Monnier, Axel Boursier & Annabelle Seoane (eds.), *Cyberhate in the context of migrations*, 63–81. Cham, Switzerland: Palgrave Macmillan/Springer.

Assimakopoulos, Stavros, Fabienne H. Baider & Sharon Millar (eds.). 2017. *Online hate speech in the European Union: A discourse analytic perspective*. Cham, Switzerland: Springer.

Austin, John L. 1962. *How to do things with words*. Oxford: Oxford University Press.

Baider, Fabienne H. 2022. Covert hate speech, conspiracy theory and anti-Semitism: Linguistic analysis versus legal judgement. *International Journal for the Semiotics of Law* 35. 1–25.

Baider, Fabienne H., Anna Constantinou & Anastasia Petrou. 2017. Metaphors related to othering the non-natives. In Stavros Assimakopoulos, Fabienne H. Baider & Sharon Millar (eds.), *Online hate speech in the European Union: A discourse analytic perspective*, 38–42. Cham, Switzerland: Springer.
Becker, Matthias J. 2020. Antisemitism on the internet: An underestimated challenge requiring research-based action. *Justice* 64. 32–40.
Becker, Matthias J. 2021. *Antisemitism in reader comments: Analogies for reckoning with the past*. Cham, Switzerland: Palgrave Macmillan/Springer.
Beyssade, Claire & Jean-Marie Marandin. 2006. From complex to simple speech acts: A bidimensional analysis of illocutionary forces. In David Schlangen & Raquel Fernández (eds.), *Brandial '06: Proceedings of the 10th Workshop on the Semantics and Pragmatics of Dialogue (SemDial-10), September 11–13, 2006, Potsdam, Germany*, 42–49. Potsdam: Universitätsverlag Potsdam.
Brandenburg v. Ohio. 1969. Supreme Court of the United States. https://supreme.justia.com/cases (accessed 21 January 2023).
Brenner, Jennifer L. 2002. True threats a more appropriate standard for analysing first amendment protection and free speech when violence is perpetrated over the internet. *North Dakota Law Review* 98(4). 753–784.
Brown, Alexander. 2017. What is hate speech? Part 1: The myth of hate. *Law and Philosophy* 36(4). 1–50.
Brown, Gillian & George Yule. 1983. *Discourse analysis*. Cambridge: Cambridge University Press.
Canadian Criminal Code. 1985. https://laws-lois.justice.gc.ca (accessed 27 August 2022).
Cannon, Walter Bradford. [1915] 1925. *Bodily changes in pain, hunger, fear and rage: An account of recent researches into the function of emotional excitement*. New York: D. Appleton and Company.
Constitution Annotated. Analysis and interpretation of the US Constitution. Amdt1.2.3.2.1. Fighting words, hostile audiences and true threats: Overview. https://constitution.congress.gov (accessed 14 September 2022).
Council of Europe. No Hate Speech Movement. https://www.coe.int/en (accessed 20 September 2022).
Criminal Code 2016. Ministry of Justice and Technical Secretariat (ed.). Clinter Traducciones e Interpretaciones (trans.) https://www.mjusticia.gob.es (accessed 20 September 2022).
Delgado, Richard & Jean Stefancic. 2018. *Must we defend Nazis? Why the First Amendment should not protect hate speech and white supremacy*. New York: New York University Press.
Delgado, Richard. [1993] 2018. Words that wound: A tort action for racial insults, epithets, and name-calling. In Mary J. Matsuda, Charles R. Lawrence III, Richard Delgado and Kimberlé Crenshaw (eds.), *Words that wound. Critical race theory, assaultive speech and the First amendment*, 89–110. New York/London: Routledge.
Grice, H. Paul. 1975. Logic and conversation. In Peter Cole & Jerry L. Morgan (eds.), *Syntax and semantics 3: Speech acts*, 41–58. New York: Academic Press.
Guillén-Nieto, Victoria. 2022. Language as evidence in workplace harassment. *CORELA. Cognition, Représentation, Langage*. HS-36. Open Edition Journal.
Guillén-Nieto, Victoria. 2023. *Hate speech: Linguistic perspectives*. Berlin/Boston: De Gruyter Mouton.
International Convention on the Elimination of All Forms of Racial Discrimination. 1965. https://www.ohchr.org (accessed 5 November 2022).
International Covenant on Civil and Political Rights. 1966. https://www.ohchr.org (accessed 5 November 2022).
Ivanou, Uladzislau. 2017. Stereotyping vulnerable groups. In Stavros Assimakopoulos, Fabienne H. Baider & Sharon Millar (eds.), *Online hate speech in the European Union: A discourse analytic perspective*, 29–32. Cham, Switzerland: Springer.

Jersild v. Denmark. Application No. 15890/89. European Court of Human Rights. Judgment. Strasbourg. 23 September 1994. Official English version of the judgment.

Kopytowska, Monika, Julita Woźniak & Lukasz Grabowski. 2017. From 'patriotism' to hate: Axiological urgency in online comments related to refugees. In Stavros Assimakopoulos, Fabienne H. Baider & Sharon Millar (eds.), *Online hate speech in the European Union: A discourse analytic perspective*, 32–38. Cham, Switzerland: Springer.

Leymann, Heinz. 1990. Mobbing and psychological terror at work. *Violence and Victims* 5(2). 119–126.

Liao, Meizhen & Sun Yadi. 2017. Cooperation in Chinese courtroom discourse. In Janet Giltrow & Dieter A. Stein (eds.), *The pragmatic turn in law: Inference and interpretation*, 57–82. Berlin/Boston: De Gruyter Mouton.

Malone, Barry. 2015. Why Al Jazeera will not say Mediterranean "migrants". *AlJazeera* https://www.aljazeera.com/features/2015/8/20/whyal-jazeera-will-not-say-mediterranean-migrants (accessed 21 April 2023).

Monnier, Angeliki, Axel Boursier & Annabelle Seoane (eds.). 2022a. *Cyberhate in the context of migrations*. Cham, Switzerland: Palgrave Macmillan/Springer.

Monnier, Angeliki, Axel Boursier & Annabelle Seoane. 2022b. Anti-migrant hate speech as a symptom of a representation crisis. In Angeliki Monnier, Axel Boursier & Annabelle Seoane (eds.), *Cyberhate in the context of migrations*, 17. Cham, Switzerland: Palgrave Macmillan/Springer.

Paasch-Colberg, Sünje, Joachim Trebbe, Christian Strippel & Martin Emmer. 2022. Insults, criminalisation, and calls for violence: Forms of hate speech and offensive language in German user comments on immigration. In Angeliki Monnier, Axel Boursier & Annabelle Seoane (eds.), *Cyberhate in the context of migrations*, 137–163. Cham, Switzerland: Palgrave Macmillan/Springer.

Public Order Act 1986. https://www.legislation.gov.uk (accessed 5 November 2022).

Rabat Plan of Action. 2012. United Nations. Human rights. Office of the High Commissioner. https://www.ohchr.org (accessed 18 December 2022).

Racial Vilification Act. 2004. https://www.legislation.wa.gov.au (accessed 13 November 2022).

Report of the United Nations High Commissioner for Human Rights on the Expert Workshops on the Prohibition of Incitement to National, Racial or Religious Hatred. 2013. https://www.ohchr.org (accessed 18 December 2022).

Ruzaite, Jurate. 2018. In search of hate speech in Lithuanian public discourse: A corpus-assisted analysis of online comments. *Lodz Papers in Pragmatics* 14(1). 93–116.

Schröter, Melani. 2022. "Everything goes against the German here!" Self-victimising discourse in comments on migration-related posts on the alternative für Deutschland Facebook page. In Angeliki Monnier, Axel Boursier & Annabelle Seoane (eds.), *Cyberhate in the context of migrations*, 83–114. Cham, Switzerland: Palgrave Macmillan/Springer.

Searle, John R. 1969. *Speech acts*. Cambridge, UK: Cambridge University Press.

Searle, John R. 1975. A classification of illocutionary acts. *Language and Society* 5(1). 1–23.

Searle, John R. 1979. *Expression and meaning: Studies in the theory of speech acts*. Cambridge, UK: Cambridge University Press.

Spanish Criminal Code. 2013. Ministerio de Justicia—Secretaría General Técnica. Trad. Clinter. Colección Traducción del derecho español. https://www.legislationline.org (accessed 18 December 2022).

Stein, Dieter. 2022. Mobbing as a genre and cause for legal action? Linguistic prolegomena for a legal issue. *CORELA: Cognition, Représentation, Langage* HS-36.

Stollznow, Karen. 2017. *The language of discrimination*. München: LINCOM GmbH.

Strani, Katerina & Anna Szczepaniak-Kozak. 2022. Online hate speech in the UK and Poland: A case-study of online reactions to the killing of Arkadiusz Józwik. In Angeliki Monnier, Axel Boursier &

Annabelle Seoane (eds.), *Cyberhate in the context of migrations*, 21–61. Cham, Switzerland: Palgrave Macmillan/Springer.

Tsesis, Alexander. 2002. *Destructive messages: How hate speech paves the way for harmful social movements*. New York: New York University Press.

US Constitutional Law. Amdt 1.2.3.2.1. Fighting Words, Hostile Audiences and True Threats: Overview. https://constitution.congress.gov (accessed 14 September 2022).

van Dijk, Teun A. [1977] 1992. *Text and context: Explorations in the semantics and pragmatics of discourse*. London: Longman.

Vella Muskat, Rebecca & Stavros Assimakopoulos. 2017. The implicit dimension of discriminatory discourse. In Stavros Assimakopoulos, Fabienne H. Baider & Sharon Millar (eds.), *Online hate speech in the European Union: A discourse analytic perspective*, 42–46. Cham, Switzerland: Springer.

Antonio Doval Pais
Chapter 9
When the wording of the law is not enough: Hate speech crimes in Spain

1 Introduction

The interpretation of so-called "hate speech crimes" in Spanish law presents numerous challenges, both from the point of view of the scope of their prohibitions and of their justification for the criminal punishment of excesses in the exercise of freedom of expression. The former is directly linked to the use of language, but language also plays a fundamental role in the latter. The kind of difficulties that arise in Spain are similar to those found in other countries, like Germany, whose regulation on this matter partly inspired the Spanish Criminal Code, France or the United Kingdom, as well as in other countries.

An analysis of these problems will constitute the main part of this paper. Before focusing on them, however, I will consider three background aspects of the issue: the specific sociodemographic context in which these crimes first appeared in Spanish law, the influence of European law in the punishment of crimes motivated by xenophobia or racism, and the type of legal instruments Spanish criminal law provides to protect foreigners and nationals of different races. It is with respect to this last and most important aspect that I will focus on the analysis of the problems posed by the language used in the law, and will discuss some solutions to overcome them.

As is well known, law is composed of language. Laws, in the same way as judicial sentences and legal acts, are expressed through natural language, that is, not through formalised language, even when highly technical legal terms are used. It is true that the law benefits when it is expressed in ordinary language (and is therefore accessible to non-specialists), but it is also limited by the endemic shortcomings of ordinary language: the ambiguity of words and syntactic structures, the vagueness of concepts (Atienza 2017: 16–22), along with other

---

Note: This paper originated in the research project: "Inequality imprints in the criminal justice system: critical analysis of the necessity, opportunity and dangerousness reasons for the difference (ÆQUALITAS)" (ref. RTI2018-096398-B-I00) funded by the Spanish Ministry of Science and Innovation.

Antonio Doval Pais, University of Alicante

problems that in practice lead to disagreements in interpretation, unequal application of laws by judges and courts or, for example, real situations of anomie (normlessness), as well as numerous other difficulties. However, the interpretation of the law does not depend exclusively on the meaning of its wording.

Moreover, the law is expressed through ordinary language on the basis of logical, systematic and teleological criteria. Logical criteria imply that any contradictions existing between rules themselves, between one principle and another, as well as between rules and principles, are inadmissible. Systematic criteria enrich the expression of the law by providing a framework for interpretation through the ordering of crimes under specific headings in the Criminal Code (e.g. "crimes against liberty", "crimes against property"). Teleological criteria enable a particular interpretation of the law in the light of the aims pursued by a given regulation. Furthermore, constitutional principles provide axiological criteria that should guide the interpretation and help resolve conflicts that arise, criteria such as the principle of equality, legality, proportionality, among other principles. All of these are essential when dealing with the challenges posed by hate crimes.

2 The origins of the criminal protection of foreigners or citizens of different races in Spain

The phenomenon of mass migration has significantly changed the human landscape of countries that have traditionally been the origin, rather than the destination, of migratory flows. Spain, like Italy and Greece, went from being a country of emigrants during the 1950s and 1960s to being, especially since the 1990s, a destination for migrants, mainly from Latin American, North African and Eastern European countries.

In Spain the need to deal politically with the phenomenon of immigration has, therefore, only been a concern for the last decades. Spain has gone from implementing measures aimed at the emigration of its citizens to other (mainly European) countries (like those established by the Spanish Institute of Emigration in the 1950s to organise the departure of Spaniards), to applying measures aimed at regulating immigration into Spain. The first decision took the form of the 7/1985 *Law on the Rights and Freedoms of Foreigners in Spain*. Although, due to the characteristics of the foreign population then residing in Spain, no problems of *xenophobia* had ever arisen, the same cannot be said of *racism*—in the 1970s alarming attacks on the Roma community had been recorded in several towns in Andalusia (Río Ruiz 2003).

It should be made clear that in Spain, from the late 1960s onwards, there was an increase in the arrival of pensioners from other European states, because their pensions had greater purchasing power in Spain, while they also benefited from a pleasant climate in a reasonably safe environment. This was, however, a very particular type of immigration. For example, in the province of Alicante there was a steady growth of residential estates populated by elderly people of the same non-Spanish nationalities. Because these estates are usually quite far from urban centres, these migrant pensioners had little contact with the local population, and were almost invisible to the Spanish general public.[1]

This profile of Spain's migrant population changed radically in the 1990s, when Spain became the European country with the third highest rate of immigration growth, due to the progressive arrival of so-called "immigrants from the south" (Pereda et al. 2003). These persons arrived in Spain in very precarious conditions, many with years of suffering behind them, having come from impoverished countries that were often politically or socially unstable.[2] These immigrants were no longer well-off elderly Europeans, who generally led comfortable lives, but rather younger people with economic problems who might subsequently be seen on the streets asking for work, food or money. Spanish legislation regarding foreigners does not generally allow for the regularisation of those who are in Spain illegally. Exceptionally, only after three years of *de facto* residence can a foreign subject achieve regularisation, if she or he has a work contract and no criminal record. During the initial three-year period following arrival, all foreigners are forced to depend on the help of others or on the informal economy.

The problems with immigration that led to the need for the intervention of criminal law arose when the first serious racist, xenophobic attack occurred: the fatal shooting of the Dominican Lucrecia Pérez, in Madrid, by individuals from the political far-right. Pérez's murder brought to light the urgent need for punitive measures for hate crimes, which until then had been completely unnecessary.[3]

The first criminal provision aimed at preventing racist attacks, and/or attacks on immigrants, dates from 1995. It introduced an aggravating circumstance for the commission of offences against persons or property on racist, xenophobic, anti-Semitic or other grounds relating to the ethnic or national origin, ideology, religion or beliefs of the victim. It also added the offence of provoking discrimination against persons or

---

[1] For more details on the characteristics of this immigration, see Pereda, Actis and De Prada (2003).
[2] Regarding the conditions of these kind of migration see Chapter 8 in this volume.
[3] On the dangerousness attributed to the migrant ("the dangerous outsider") see Chapter 2 in this volume.

groups on grounds related to their racial, ethnic or national origin or to their ideology, religion or beliefs, the class of offence now known in Spain as *hate crime*.[4] The urgent need for these provisions was mentioned shortly after Pérez's murder, in the Explanatory Memorandum to the law that included in the Criminal Code both the aggravating circumstance and the offence, stating: "The proliferation of episodes of racist and anti-Semitic violence perpetrated under the flags and symbols of Nazi ideology obliges democratic states to take decisive action to combat it" (LO 4/1995).

Thus, since the mid-1980s Spain has had rules regulating the conditions of immigrants, and since the mid-1990s, there have been criminal rules specifically aimed at their protection. From this point onward, Spain's immigration policies have been broadly characterised by three guidelines: that immigrants should not leave their countries of origin, should not enter Spain and, if in Spain, should leave (Lara 2008: 91–92). These decisions have been enshrined in agreements with countries from which the flows originate or through which they pass on their journey to Spain, as well as in various laws aimed at regulating the conditions of entry and stay of non-EU foreigners. They have even been reflected in the Spanish Criminal Code, which enables a judge to deport foreign citizens convicted of crimes. However, other measures, both of a civil and criminal nature, have also been adopted specifically for the protection of immigrants.

3 Crimes motivated by xenophobia or racism: The influence of European law

The use of criminal law for the prosecution of xenophobic or racist hate crimes has been directly promoted, firstly, by the Council of Europe whose monitoring body in this area, the European Commission against Racism and Intolerance (ECRI), adopted a number of recommendations on the criminal prosecution of hate crimes, in order to fight racism, xenophobia, anti-Semitism and intolerance.[5] Secondly, it has been promoted by the European Court of Human Rights (ECHR), responsible for ensuring respect for the rights set out in the European Convention on Human Rights, obliging

---

[4] Organic Law (Ley Orgánica) 4/1995 Amending the Criminal Code for Punishing the Direct and Public Incitement to Commit Genocide (hereinafter, LO 4/1995).
[5] General Policy Recommendation No. 1 of 1996 on Combating Racism, Xenophobia, Antisemitism and Intolerance, rec. A); General Policy Recommendation No. 7 of 2002 on National Legislation to Combat Racism, paragraph IV; and Racial Discrimination and General Policy Recommendation No. 15 of 2015 on Combating Hate Speech, rec. 10. On the latter provision, see Chapter 8 in this volume.

member states to include hate crimes in their criminal codes.[6] Thirdly, in the framework of the European Union, although the non-discrimination directives do not oblige EU Member States to use criminal law to prosecute hate speech, a Framework Decision by the European Council does oblige all EU Member States to provide for criminal sanctions in relation to incitement to violence or hatred based on race, colour, descent, religion or belief, national or ethnic origin. The ECRI framework also refers, in this sense, to the dissemination of racist or xenophobic material and the condoning, denial or trivialisation of genocide, war crimes or crimes against humanity. Member States are also obliged to consider racist or xenophobic intent as an aggravating circumstance.[7]

Although the ECRI recommends limiting the use of criminal law to punish the most serious and extreme cases of incitement to hatred,[8] there is a wide margin for states to specify the criminal effects of these behaviours, and the ECHR does not consider an explicit call to violence as essential in order to identify an act as a hate crime.[9] Hereafter I will explain how laws relating to hate crimes motivated by racism and xenophobia have been implemented in Spain, and the problems that the formulation of the corresponding provisions have caused in practice.

4 Current protection of foreigners and national citizens of different races in Spanish criminal law

With regard to the protection of foreigners or of national citizens of different races, since 1995 Spanish criminal law has maintained the specific qualifying circumstance of acting on discriminatory grounds and the offence of provoking discrimination. Yet, through successive reforms, it has considerably extended both the scope of the qualifying circumstance of discrimination (broadening the type of offences susceptible to aggravation and also the protected groups) and the scope of the so-called hate crime (which after 2015[10] no longer spoke of provoking, but of fostering, promoting or inciting discrimination). It also incorporated various offences, designed to protect

---

6 ECHR Angelova and Iliev v. Bulgaria, Application No. 55523/00, 26th July 2007 [Section V].
7 Council Framework Decision 2008/913/JHA of 28th November 2008 on Combating Certain Forms and Expressions of Racism and Xenophobia by means of Criminal Law, OJ L 328, 6th December 2008.
8 ECRI General Policy Recommendation No. 15.
9 See for instance ECHR Féret v. Belgium, Application No. 15615/07, 16th July 2009.
10 LO 1/2015 amending LO 10/1995 of the Criminal Code.

foreigners or national citizens of different races in specific areas. For this purpose, threats of harm aimed at causing fear among an ethnic, cultural or religious group, or any other group of persons, offences of human trafficking, abusive working conditions, and the denial of the right to have access to public services, were included. Membership in unlawful associations which promote racial discrimination was also added as an offence in the Criminal Code.[11]

Clearly, not all of these crimes should be considered hate crimes, nor are they all committed through verbal attacks (those which characterise hate speech), nor is the inclusion of these crimes only aimed at protecting immigrants. Some of the crimes can affect national citizens (for instance human trafficking, or even hate crimes themselves[12]) or cannot be committed in spoken or written form (e.g. again, human trafficking, or crimes against workers' rights).

Because this paper deals exclusively with hate crimes, I would like to make the following observations.

To start, I must underline the fact that the category of "hate crimes" is more criminological than legal, not only in Spain but also in other countries. In the international context, the heading "hate crime" does not generally appear in criminal laws. Hate crimes are simply considered those behaviours carried out for xenophobic or racist (or other discriminatory) reasons.

I would also like to make the general observation that, as I have already pointed out, like in other countries, conducts carried out for such motives can have two different effects. First, they can lead to an aggravation of criminal liability. Second, acting on the grounds of hatred can give rise to criminal offences, or can transform a neutral conduct, which is in principle not criminal, into a criminal offence. In Spanish criminal law these two paths form an anti-discrimination system that combines the Continental European model, based on specific hate speech crimes, with the Anglo-Saxon model characterised by the qualification of basic crimes.

A final general observation is that these two effects cannot be combined to further aggravate criminal responsibility, because the *non bis in idem* principle

---

11 Respectively, with reference to all the above offences, sections 170.1, 177 bis, 312–314, 511–512 and 515.2ª of the Spanish Criminal Code. Offences "against the rights of foreign nationals" (sec. 318 bis) should be added to this group. As well as the offence of female genital mutilation (sec. 149.2), the presence of which in our law is also the result of multiculturalism. In this case, the aim is to protect foreign minors against foreigners from countries where this practice is established.

12 In fact, the data provided in Spain by the Ministry of Interior show that a large number of those who claim to be victims of the reported incidents of hate crime had Spanish nationality (250 out of a total of 614, according to data from 2020). See Annual Series. Hate crimes. Victimisations by cause of hate crimes, available at: https://estadisticasdecriminalidad.ses.mir.es (accessed 9 February 2023).

("prohibition of double jeopardy") prevents it. In other words, the penalty for the specific crime of racial or xenophobic hatred cannot be enhanced because the offender acted for racist or discriminatory reasons (as this fact is already implicit in the crime itself).

I will now examine some of the characteristics which define each of these two preventive and punitive instruments.

4.1 The aggravating circumstance of acting for racist or xenophobic motives

Among the reasons the Spanish Criminal Code contemplates for increasing a penalty is the committing of a crime "for racist, anti-Semitic or anti-Roma reasons, or another kind of discrimination related to ideology, religion or belief of the victim, ethnicity, race or nation to which he belongs, his sex, age, sexual or gender orientation or identity, reasons related to gender, aporophobia or social exclusion, illness suffered or disability" (section 22.4).[13]

This regulation refers to motives, which, being subjective elements, lead to very imprecise interpretations. Regarding the different types of discriminatory grounds, Spanish law provides a *numerus clausus* list, which is growing as a result of recent reforms.[14] By virtue of the principle of legality, the aggravating circumstance cannot be applied to other motives which are not contemplated (prohibition of analogy *in malam partem*). Nevertheless, the grounds here should not require any further expansion, since a wide variety of broad terms are already used, some of which are difficult to distinguish from others, such as "racist reasons" and "other discrimination based on race". The references of direct interest to us here are the following: "racist, anti-Semitic or anti-Roma reasons" and "other kind of

---

13 This is the English version provided by the Ministry of Justice of the Kingdom of Spain, available at: https://www.mjusticia.gob.es (accessed 9 February 2023). It has been updated by the author in accordance with both the Organic Law 8/2021 on the Full Protection of Children and Adolescents against Violence (hereinafter, LO 8/2021) to include references to *age*, *gender* and *aporophobia* (phobia of poor or disadvantaged people) or social exclusion and with the Organic Law 6/2022 on the Full Protection for Equal Treatment and Non-Discrimination (hereinafter, LO 6/2022) to include anti-Roma grounds. On *aporophobia* and its relation to fear of poor people ("the dangerous poor"), see Chapter 2 in this volume.

14 As noted before (footnote 13), by the LO 8/2021 and the LO 6/2022. Earlier reforms had also incorporated new grounds in 2010 (sexual identity) and 2015 (gender-related reasons). It should be noted that the amendment including anti-Roma grounds affects not only the aggravating circumstance of section 22.4 but also the offences contemplated in section 510 of the Spanish Criminal Code.

discrimination" related to the victim's "ethnicity, race or the nation to which he belongs". However, it is true that "religion or belief of the victim" could also constitute grounds associated with racism or xenophobia (Borja Jiménez 1999: 259–263).

In the rulings of our courts the aggravating circumstance is applied most frequently in crimes of homicide and injury (Hortal Ibarra 2012: 33). Nevertheless, it is applicable to any offence committed, regardless of its seriousness, for example whether someone is killed, beaten, threatened or robbed. This is also the case, of course, when an offence is committed through language (i.e. threats, intimidation, offences against honour or certain offences against privacy, among others). As mentioned above, this circumstance does not apply when it is inherent to the crime committed, as in the cases of incitement to racial hatred (sec. 510, which will be discussed below), or denial of the right to public services (sec. 511), in order to avoid a *bis in idem*. I mention it again here because this is sometimes disputed; for example, in the case of insults motivated by racism or xenophobia. The issue, therefore, goes beyond legal language and here requires a teleological argument so as to answer the following question: Do both provisions aim to protect people against the same kind of harm?

The legal reference to "motives" is generally interpreted as a reference to an *impulse*, or the reason for the offence. The motive (from the Latin, *motus, motus*: movement) must move the offender, in the sense that it must impel the offence or must be the "spring" or "originator" of the actions (Sistare 1987: 305). Motive is, therefore, a decisive element. Moreover, the point at which it is present is key, as it is not enough for the motive to appear at just any point in the course of the action (Cuerda Arnau 1996: 242). For instance, if the aggression started for another reason and, afterwards, during the aggression, racist expressions were uttered, this is insufficient to find racism as a motive. This was apparently the case in a confrontation over a possible mobile phone video recording in Spain in 2021, which resulted in the murder of Samuel, a 24-year-old homosexual. According to news reports, before the assault the offender stated "Stop recording us if you don't want me to kill you", and subsequently called Samuel *maricón* (a homophobic slur in Spanish).[15]

Yet, a certain extensional vagueness (Atienza 2017: 19) can be perceived in the legal expression "for reasons", which could imply that in the above case the offence may have been committed "for racist or xenophobic reasons". But the legal reference apparently intends to focus only on those reasons that directly

---

15 Regarding this case, see the July 5 2021 article by Marta Borraz in Spain's elDiario.es: "El crimen de Samuel desata una oleada de protestas contra la violencia que sufren las personas LGTBI" ('Samuel's murder sparks wave of protests against violence against LGTBI people'). https://www.eldiario.es/ (accessed 5 March 2023).

trigger the offence, which does not preclude the possibility that there may be an alternative explanation for the aggressive reaction compatible with a racist or xenophobic impulse. For example, this is the case when a reaction of self-defence is more extreme because the aggressor is of a different race and, therefore, the reaction involves an intensive excess (i.e. a more aggressive reaction than needed) in the use of self-defence. The admissibility of the qualifying circumstance is questionable when the xenophobic or racist motive has not itself triggered the crime (for example, if the offender has already decided to commit a fraudulent offence), but has only determined the choice of victim (the offender selected the victim of the fraud because, as an immigrant recently arrived in Spain, he does not know our language and is an easier target).[16]

As the courts state, motives can only be inferred from the indications of all the elements that constitute the context of the act (attitude of the aggressor prior to and at the time of the act, outward signs of the aggressor's beliefs, such as tattoos or the way he/she is dressed, characteristics of the victim, and other elements of this kind). These indications must be concretely verified, and they must also make it possible to reasonably infer motives, in accordance with the rules of rationality and common experience. The Spanish courts frequently refer to the fact that the discriminatory motive must be proved, among other indications, by means of offensive expressions directed at the victim. In this case, it is necessary to examine not only what is said, but also how it is said (Solan & Tiersma 2018: 149). However, it is obvious that no verbal expression is, in general, necessary for the motive to be present (Hortal Ibarra 2012: 52). For this reason, it is essential to examine the context in which the event took place.

Motives lie in the perpetrator's inner self, and whether criminal law can take into account the motives of actions has been much debated, because liberal criminal law is a criminal law of facts, and not a law of ways of being, beliefs, thoughts or consciences. Nevertheless, disfavoured motivation is not only found in hate laws, but also in other areas of criminal law (Steiker 1999: 1861–1872).

The discussion regarding the justification of penalty enhancement for acting on certain motives has traditionally offered either an explanation based on the blameworthiness of the perpetrator or, alternatively, on the wrongfulness of the act (Laurenzo Copello 1996: 274–288). Yet, some authors have positioned themselves very strongly against any reasonable justification for this aggravation of penalties.

A significant sector of the doctrine considers that the increase in the penalty for motives responds to the fact that the conduct deserves greater reproach (is

---

[16] I take this example from Laurenzo Copello (1996: 273).

more blameworthy) because it is carried out for abject motives. However, another large group of authors see the actions stemming from these motives as constituting a greater harm than actions carried out without them: the perpetrator is punished more severely because his conduct is more harmful from an objective point of view. From this perspective, it deserves a higher penalty on the grounds of proportionality between the weight of the harm and that of the punishment.

Each of these positions requires further justification. On the one hand, authors who argue that individuals who are motivated by socially unacceptable reasons are guiltier, and thus deserve to be punished more severely, do so on the grounds that the offences are more morally reprehensible. On the other hand, authors who argue that a crime motivated by bias produces a greater harm consider that this is either because of the additional harms to equality (non-discrimination), or because the peace and security of other people with similar characteristics to the victim are adversely impacted.

There has been a great deal of intense discussion on this point, and the discussion is not limited to these two alternatives of the so-called wrongfulness-culpability paradigm. As I have pointed out, a state can impose harsher sanctions through antidiscrimination criminal law as a way to provide fair protection of vulnerable populations (Harel & Parchomovsky 1999: 523–529). Nevertheless, this is not merely an academic question, but one that has practical consequences for the interpretation and application of the law in real cases. On the one hand, standards of evidence could be simplified, both if the qualifying circumstance were considered to respond to reasons of depravity on the part of the criminal, or if it were based on the presumption of greater harm because the offence was also an attack on the right to equality.[17] According to either of these interpretations, it is only the fact that the defendant carried out the criminal act for an abject motive that needs to be proved before the judge, because that would constitute sufficient proof that the offence was a more blameworthy action and/or an attack against equality. On the other hand, if the reason for the qualifying circumstance is the impact on the peace or security of other members of the victim group, proof of this will be required. Therefore, once again, the wording of the law alone is not enough to provide a solution. Careful interpretation is required in order to clarify the meaning of the law, and such interpretation must be rooted in the reasons behind the existence of these criminal rules (teleological reasons).

---

[17] The justification based on greater blameworthiness could also explain the punishment when the offender is mistaken about the concurrence of the discriminatory grounds in the victim, since regardless of whether he was right or wrong the offender acted for reprehensible motives (Laurenzo Copello 1996: 277).

One final important point regarding the scope of the qualifying circumstance relates to the characteristics of the victim. The Criminal Code section does not expressly require that the offence be directed against a person who actually belongs to another race, ethnicity or nation. The question of whether the victim must possess any of these characteristics, or must only be selected for a discriminatory reason (e.g. because of his or her foreign appearance), has been discussed in practice, and the Spanish Supreme Court has maintained various criteria in its rulings.[18] This is why the LO 8/2021 penal reform has resolved the doubt, by adding the following clarification: "regardless of whether the person affected by this conduct in fact possesses such conditions or circumstances". Thus, anyone who acts even in the mistaken belief that the victim is of a different race is indeed acting on racist grounds.

The same question regarding the characteristics of the victim arises when he or she is a third party attacked for defending another person who is a victim of racist aggression, or because of his or her real or supposed links with individuals possessing any of these characteristics. In this case too, the aggressor would be acting against the defender "for racist motives". On the one hand, the law speaks of "committing the offence *for* racist motives", in which case the motive need not be linked to the victim personally. It is sufficient that the perpetrator acts "because of" one of the motives contemplated, even if the main attack (that is, the offence: let us not forget that we are talking about an aggravating circumstance) is not on the person discriminated against. The section, nevertheless, also refers to "other discrimination *based on* the ethnicity, race or nation to which [*scil.* the victim] belongs", which would prevent the application of the aggravating circumstance if, for example, at the moment of striking the victim the aggressor had said (and I quote): "this is what you get for being the Jew's son".[19]

Consequently, it is not clear in the law how far the reason for the discrimination goes. Here the problem stems from the use of language, which generates a syntactic ambiguity and raises doubts about the model followed by the law to punish such conduct.

On this issue, although some tribunals have required that the group to which the victim belongs be either a vulnerable or a minority group, this is not required in Spanish criminal law. Therefore, the wording of the law makes it possible for the aggravating circumstance to apply in cases of hatred against, for instance, the

---

[18] In fact, this was one of the aspects referred to by the State Public Prosecutor's Office in its Guidelines for interpreting hate crimes under section 510 of the Criminal Code (2019), to be followed by each member of the Public Prosecution Service.
[19] As was concluded in the Judgement of the Audiencia Provincial (hereinafter referred to as SAP, equivalent to the English Crown Court) of Burgos 420/2001, 20th October.

police or white people.[20] However, there are other reasons why such an interpretation should be excluded, based on the aim that can be deduced of the penal intervention in this area: that it is intended to protect members of vulnerable, marginalised or racialised populations from racism and xenophobia (Laurenzo Copello 2019: 464–467).

A second way in which acts motivated by racism or xenophobia can have criminal consequences is found in what are known in Spain as "hate speech crimes".

4.2 Hate speech crimes

Actions motivated by racism or xenophobia can not only increase the penalty of an offence, as we have just seen, but can also incur criminal consequences.

The Spanish Criminal Code provides sanctions for the following types of conduct: public incitement to hatred, hostility, discrimination or violence; acts related to writings or other material or media which could promote incitement to hatred; the extolling of genocide and other similar offences; degradation of groups or persons; the extolling or justification of racist, xenophobic, homophobic offences by others, and unlawful association for the incitement to hatred, hostility, discrimination or violence.

All of these behaviours are characterised under the law by the fact that the action is driven by racist, xenophobic or anti-Semitic motives,[21] or any other kind of discrimination related to ideology, religion, beliefs, family situation, ethnicity, race or nation, national origin, sex, sexual orientation or identity, gender, illness or disability. These grounds are almost the same as those we have just seen, which aggravate liability in general. The difference is that here we are dealing exclusively with offences committed solely by means of speaking, writing, or via any other communicative material or media. Below, I quote the most relevant paragraphs of section 510, the provision which contains the abovementioned conducts ("archetypical of hate speech", in the words of the Spanish Supreme Court), to support and clarify my explanation. This provision has been strongly criticised by the doctrine for a number of reasons, some of which will be discussed below.[22]

---

20 Regarding white people, one example is the widely known case of Todd Mitchell, which led to the US Supreme Court ruling in the case of *Wisconsin v. Mitchell*, 508 US 476 (1993).
21 The first five types of conduct also include anti-Roma grounds after LO 6/2022.
22 For further reasons for criticism see Chapter 8 in this volume.

Section 510

1. A prison sentence of one to four years and a fine of six to twelve months shall be imposed on:
   a) Those who, directly or indirectly, foster, promote or incite hatred, hostility, discrimination or violence against a group, or part thereof, or against a certain person for belonging to such a group, for reasons of racism, anti-Semitism, *anti-Roma* or for other reasons related to ideology, religion or beliefs, family circumstances, the fact that the members belong to an ethnicity, race or nation, national origin, gender, sexual orientation or identity, or due to gender, *aporophobia*, illness or disability;
   b) Those who produce, prepare, possess with the purpose of distributing, provide third parties access to, distribute, publish or sell documents or any other type of material or medium that, due to the content thereof, are liable to directly or indirectly foster, promote or incite hatred, hostility, discrimination or violence against a group, or part thereof, or against a certain person for belonging to such a group, for reasons of racism, anti-Semitism, anti-Roma or for other reasons related to ideology, religion or beliefs, family circumstances, the fact that the members belong to an ethnicity, race or nation, national origin, gender, sexual orientation or identity, or due to gender, illness or disability;
   c) Those who publicly deny, seriously trivialise or extol the crimes of genocide, crimes against humanity or against persons and property protected in the event of armed conflict, or who extol the perpetrators thereof, if committed against a group or part thereof, or against a certain person for belonging to such a group, for reasons of racism, anti-Semitism, anti-Roma or for other reasons related to ideology, religion or beliefs, family circumstances, the fact that the members belong to an ethnicity, race or nation, national origin, gender, sexual orientation or identity, or due to gender, illness or disability, if such conduct promotes or encourages a climate of violence, hostility, hatred or discrimination against such individuals.
2. A prison sentence of six months to two years and a fine of six to twelve months shall be imposed on:
   a) Those who harm the dignity of persons through actions that entail humiliation, disregard or discredit of any of the groups outlined in the preceding Section, or of a part thereof, or against a certain person for belonging to such a group, for reasons of racism, anti-Semitism, anti-Roma or for other reasons related to ideology, religion or beliefs, family circumstances, the fact that the members belong to an ethnicity, race or nation, national origin, gender, sexual orientation or identity, or due to gender, illness or disability, or who produce, prepare, possess with the purpose of distributing, provide third parties access to, distribute, publish or sell documents or any other type of material or medium that, due to the content thereof, are liable to harm the dignity of persons because they represent serious humiliation, disregard or discredit of any of the aforementioned groups, or part thereof, or of a certain person for belonging to such a group;
   b) Those who, through any form of public expression or distribution, extol or justify the criminal offences committed against a group, or part thereof, or against a certain person for belonging to such a group, for reasons of racism, anti-Semitism, anti-Roma or for other reasons related to ideology, religion or beliefs, family circumstances, the fact that the members belong to an ethnicity, race or nation, national origin, gender,

> sexual orientation or identity, or due to gender, illness or disability, or those who have participated in its execution.
>
> The deeds shall be punished with a prison sentence of one to four years and a fine of six to twelve months if such conduct promotes or encourages a climate of violence, hostility, hatred or discrimination against the aforementioned groups [. . .].[23]

Specifically, the conducts referred to here can be understood from the point of view of the "harm paradigm", as they involve danger to life, physical integrity or liberty. Yet, the conducts can also be interpreted as offences that directly result in violations of dignity and equality (non-discrimination). From the perspective of the "culpability paradigm", these conducts can be seen as more blameworthy (Fuentes Osorio 2017: 9).

With regard to this debate, I would like to point out that I do not think the use of criminal law can be justified in either of the two ways stated above. In my view of the role of criminal law, only the first interpretation is legitimate. The blameworthiness justification is based on a moral reproach that raises too many interpretive ambiguities (Harel & Parchomovsky 1999: 512–514). The harm paradigm is inadequate, in my view, because not just any harm is sufficient to justify the intervention of criminal law, which should be reserved for only the most serious cases. That is to say, the classification of hate speech crimes should respond to the pressing need to prevent a "clear and present danger" to people's basic rights (life, physical integrity, freedom, privacy and others).[24]

As I have already pointed out, this debate has far-reaching consequences for the interpretation of the text of the law itself (which, as can be seen, is extremely vague) and, consequently, for the scope of the law in practice.

Next, I would like to briefly discuss a series of acts that have been considered punishable by the Spanish courts as hate speech crimes, as a way of providing some concrete examples of the relevant kinds of conduct. I will then consider some counter-examples of actions which have been deemed by the courts to be criminally irrelevant.

---

[23] This is the English version provided by the Ministry of Justice of the Kingdom of Spain in https://www.mjusticia.gob.es (accessed 9 February 2023). It has been updated by the author to include the anti-Roma grounds (in accordance with the LO 6/2022) and to include the reference to *aporophobia* (in accordance with the LO 8/2021).

[24] As concluded in the landmark case *Schenck v. United States*, 249 U.S. 47 (1919).

4.2.1 Examples of hate speech court cases

Among the acts of speech punished as criminal hate offences in Spain in recent years, I would like to present the following six. I have selected these examples because they took place in a wide variety of scenarios: in the media or on social networks (cases one and four), in the presence of many like-minded people (cases two and five), and in the presence of the victim and family members or colleagues (cases three and six).

Example one. Publication in a newspaper (of wide circulation) in Santa Cruz de Tenerife of a poem calling for the violent expulsion of African immigrants arriving at the coast of the Canary Islands.[25]

Example two. Songs sung at a concert with the following messages, referring to foreigners and Jews: "You can give yourself up for lost [. . .] we will defend the street, wild and aggressive, we will hit hard, we will die killing, camouflage yourself; this is not a game, with Jews there will be no mercy [. . .] We are going to execute you now. You are going to die. Your bodies will hang from trees and lampposts and your children will die".[26]

Example three. Statements directed against the bride and her family and friends, all Indians, at a wedding celebration: "Indian whores: they come to grab a Spaniard to get his money".[27]

Example four. Numerous messages against people of foreign origin posted on a blog, messages specifically directed at immigrants of islamic faith, Jews and people of African descent, attributing to them, among other features, either fanaticism, evil, ambition, corruption, filth or brutality, while defending the supremacy of the Aryan race and including what are known as "The Fourteen Words".[28]

Example five. Various declarations at an event held following a demonstration by far-right parties, claiming that the EU subjugates Spain completely and has brought a wave of Islamists to Spain as refugees, and that refugees are not welcome.[29]

---

25 SAP Santa Cruz de Tenerife 107/2014, 7th March.
26 Spanish Supreme Court Judgement (hereinafter referred to as STS) 675/2020, 11th January.
27 SAP Valladolid 56/2019, 8th March.
28 "We must secure the existence of our people and a future for white children". Racist slogan by David Eden Lane, member of the white separatist insurrectionist group The Order. See SAP Barcelona 392/2018, 14th May.
29 SAP Barcelona, 27th April 2018.

Example six. The statement "Go and treat monkeys in your country", uttered orally at a Venezuelan doctor in the emergency ward of a hospital, within hearing range of other health care professionals who were present.[30]

The above were either punished as crimes of incitement to hatred, hostility, discrimination or violence (sec. 510.1, a: cases one, two, four and five), or as crimes of humiliation or degradation of groups or persons (sec. 510.2, a: cases three and six). The penalties imposed were very close to the applicable minimum (generally, 6 months or 1 year imprisonment and a day fine of between 4 and 10 euros a day for 6 months). The most severely punished were cases one, two, four and five, all characterised by the extensive public diffusion of the hate speech in question.

For comparison, I have also selected three more cases in which judges understood the facts as having had no criminal relevance, for the reasons I will indicate below. Two of these cases took place in the context of political campaigns, in which freedom of speech plays a special role. This context was taken into account by the judges in cases one and three, described below. Although the message in the third was very clear, this case underlines the legal importance of racism or xenophobia as the motive that triggers or originates the crime, as the appearance of such motives at some later point during the event is inadequate. The cases were the following:

Case one. Declarations made in public interviews and in party leaflets by a mayoral candidate in Barcelona during an election campaign. In these utterances, the candidate stated that the Roma groups had come to Catalonia to commit crimes, and that the Roma collectives should be treated firmly and met with police force, that they should be arrested, imprisoned and returned to where they came from. In this case, the court concluded that the intention guiding the accused was that of denouncing the existence of citizen insecurity, not racism or xenophobia. The court therefore considered that he did not act with *animus iniuriandi* ('intent to offend'), and that the defendant did not make these statements because the Roma belonged to any particular ethnic group.[31]

Case two. The expression: "Fucking Paki", uttered during a fist-fight. As the tribunal stated, "in this context, we cannot infer any animosity towards people of that ethnic origin that can be situated within so-called hate speech".[32]

---

30 SAP Tarragona 455/2019, 8th November.
31 SAP Barcelona 713/2014, 22nd July.
32 SAP Barcelona 86/2020, 5th February.

Case three. A poster supporting the extreme right-wing political party "VOX", placed in a very central underground station in Madrid during the election campaign for the presidency of the Autonomous Community of Madrid. The poster in question compared money spent on a retirement pension with the monthly cost spent on unaccompanied foreign minors (UFM), misleading statistics (as one newspaper noted), and used the intentionally provocative image of a young man of foreign appearance wearing a black hood and a mask. Its caption was: "A UFM 4,700 euros a month. Your grandmother 426 euros pension a month". The court dismissed the case on the grounds that it did not qualify as a hate crime for the following reasons. The event took place in the context of a legitimate ideological struggle during an electoral campaign, where verbal excesses are commonplace. Moreover, the poster can be interpreted as an electoral message referring to a much larger group than only unaccompanied foreign minors: it can be read as a criticism of the insufficient size of pensions—retirement or widowhood pensions in particular— in contrast to what is spent for other purposes, such as the support of these young migrants.[33]

4.2.2 Questions regarding legal wording

The aforementioned problems of interpretation lie in the specific wording of the laws on hate crime, sometimes because of what they express, but sometimes because of what they fail to express.

The application of the legal provisions referring to hate crimes creates controversy because the wording of the laws fails to clarify which elements must be present for an act to be considered a hate crime. It will not be possible, within the scope of this chapter, to consider all the possible elements. For this reason, from here forward I will largely focus only on the most representative hate crime, the one which most often appears before the Spanish courts and the regulation of which has been most often criticised by scholars: the crime of incitement to hatred (sec. 510.1, a).

With regard to problems arising from the wording of the laws themselves, the legal terms in this field are particularly vague, especially in the case of the

---

33 See the Interlocutory Order (Auto) AP Madrid 631/2021, 1st July. The events took place in April 2021. Regarding this case, see the July 5 2021 article by Patricia Peiró in *El País*: "La justicia avala el cartel electoral de Vox contra los menores extranjeros: Son un evidente problema social y político" ('Justice endorses Vox's electoral poster against foreign minors: They are an obvious social and political problem'). https://elpais.com/ (accessed 5 March 2023).

words "to foster, to promote or to incite". This vagueness is reflected in the use of alternative terms, some of which are merely synonyms (this defective technique is common in various sections of the Spanish Criminal Code). However, not all of the terms are different ways of saying the same thing. In the case of hate speech crimes, which, as stated above, describe the conduct as "fostering, promoting or inciting", the concept of "incitement" is the most specific, while the first two words can be taken as synonyms. Thus, if all these words have the same semantic core but differ in intensity, it should have been sufficient for the law to use only the word with the widest scope, for greater linguistic economy and clarity in the normative message. That is, the law could have used only the word "to foster" (or "to promote") without adding "to incite", since anyone who incites also fosters (or promotes).

Such use of words with different scope also occurs when dealing with other criminal hate offences, such as that of providing third parties with access to, or distributing, or disseminating or selling writings, or any other type of material or media which, because of their content, are suitable for fostering, promoting or directly or indirectly inciting hatred, since such acts can be described as conducts of "fostering" or "promoting" hatred (sec. 510.1, b).

In short, the law makes it difficult both to interpret what is criminally punishable and, by providing a penalty of the same severity for conducts of varying seriousness, to ensure proportionality between the gravity of the conduct and that of the penalties.[34]

The crime of incitement to hatred is also ambiguous with regard to what is required from the agent. Thus, the description of the conduct generates uncertainty as to what exactly is being punished: Is it an action or a result of an action? In the first instance, even if no effect were achieved as a result of the action, it would be sufficient to publicly carry out an action suitable or appropriate for the dissemination of hatred, hostility, discrimination or violence for it to constitute a criminal offence. In the second, the criminal sanction would require that the subject's action demonstrably lead to greater hatred or hostility, greater discrimination or greater violence.

---

[34] It should be clarified that the text states that the penalty is of the same severity since the time interval which can be selected by the judge is restricted: imprisonment from one to four years and a daily over a period fine of six to twelve months. Within this framework, the judge can select the exact penalty that corresponds to the act, in accordance with certain rules set out in the Spanish Criminal Code. These rules make it possible to adapt the applicable penalty to the gravity of the conduct. However, there is nothing to ensure that a less serious conduct is punished with a lower penalty than that which would correspond to a more serious one.

Chapter 9 When the wording of the law is not enough: Hate speech crimes in Spain — **219**

As can be seen, this question is consequential because, to a great extent, it conditions what must be proven at trial. Adequate evidence will undoubtedly be easier to provide if the offence is understood as an offence of mere activity, one that only requires the expression, in any media or form, of language that can generate hatred, hostility, discrimination or violence. Yet, how can this doubt regarding the ambiguity of the laws be resolved? The letter of the law is not enough. It is therefore essential to use specific technical arguments provided by criminal law. For example, one possible technical argument could be made about crime prevention. From this perspective, as the Spanish Supreme Court concluded, it is inadvisable to wait until hatred is actually disseminated publicly for the conduct to be considered criminal, because of the great danger this dissemination may entail.[35] Furthermore, in my view, the word "publicly", required here by the law,[36] points, albeit subtly, towards that same conclusion, because it makes the action itself more objectively dangerous (as in examples one, two, four and five above). In other words, under this interpretation, the message does not need to be disseminated for it to "activate" the recipients' response, but it is sufficient that expressions have been made which involve incitement to hatred, hostility, discrimination or violence. This interpretation is also required by international standards and the jurisprudence of the ECHR.

Similarly, the wording of the law fails to express what jurisprudence in Spain generally considers to be an inherent requirement of this *offence*: the capacity of the conduct to incite hatred, hostility, discrimination or violence. In other words, in accordance with this interpretation, the racist or xenophobic message, by its nature and circumstances, must possess characteristics that causally enable the message to achieve that effect.[37] This requirement, which helps limit the intervention of criminal law to the most serious cases, has been defined clearly in a judicial sentence as "the potential for the act to be carried out", through its "imitative force" or through

---

35 On the danger of "hatred dissemination" as a component of the "hate speech sequence" see Chapter 8 in this volume.
36 "publicly [. . .] foster, promote or incite [. . .]" (secc. 510.1, a). The term "publicly" is not found in the English version published by the Spanish Ministry of Justice cited in footnote 23.
37 Thus, for example, STS 259/2011, 12th April, affirmed that the dissemination of ideas or doctrines are criminally prosecutable if they involve "a certain danger of generating a climate of hostility that may take the form of specific acts of violence, hatred or discrimination against certain groups or their members". Also, STS 646/2018, 14th December, stating: "Hate crime is defined in sec. 510, which does not require [. . .] the generation of a specific situation of danger, although it does require an aptitude for the generation of a situation of danger which is considered to be serious". Similarly, ATS 16th November 2020 (Case of the "Thirteen Roses"), requiring the capacity of the conduct to create the relevant danger.

messages which constitute an "effective call" to hatred, hostility, discrimination or violence.[38] The requirement responds to technical arguments specific to the validity conditions of criminal law, from the point of view of the harm model: the offence must involve damage or danger to others. In other words, innocuous conducts, which do not endanger the protected legal interest, cannot be punished. This specific construction of harmfulness requires information about all the circumstances in which the action was carried out. In practice, when one analyses a case, factors like those contained in the Rabat Plan of Action (UN) should be taken into account by the courts: context (whether they are an isolated or repeated occurrence, or whether they can be seen as compensated by other utterances); the speaker's position or status in the society (political, religious or community leader); intent; content and form (degree to which the speech was provocative and direct, as well as the form, style, nature of arguments deployed); reach or magnitude of the speech (its public nature, its magnitude and the size of its audience); and likelihood of harm (risk of harm).[39] With regard to these requirements, the Spanish Supreme Court demands an individualised investigation, with a rigorous case-by-case analysis that must examine both the specific phrases or expressions produced and the occasion and the setting in which they were uttered.[40]

In terms of the structure of the conduct, then, certain elements must be present for the speech to constitute a criminal offence. According to the law, it is required that the speech should publicly foster, promote or incite, either directly or indirectly, hatred, hostility, discrimination or violence on racist, anti-Semitic, ethnic, racial or national affiliation or national origin grounds, among others. As can be seen, the law includes foreigners as well as nationals of other races and persons characterised by belonging to an ethnic group. Here the most striking point is the specific reference to "anti-Semitic reasons", a term referring exclusively to Jews, which has been criticised as being unnecessary (this is also the case under sec. 22.4). The doctrine understands that the reason for the specification of anti-Semitic motives lies in the historical dimension of anti-Semitism in Europe (Goyena Huerta 2015: 352; Arroyo de las Heras 1997: 57), which makes anti-Semitism a powerful example of what must be prevented. The same can now be said of the anti-Roma amendment mentioned earlier.[41] These specific cases show that the words of the law not only fulfil an expressive

---

[38] SAP Barcelona 702/2018, 8th November.
[39] The Rabat Plan of Action (United Nations, 2012) sets out a six-factor test to assess the capacity of speech to create a serious risk of inciting discrimination, violence or other lawless action. https://www.ohchr.org (accessed 9 February 2023).
[40] For instance, STS 299/2011, 25th April or STS 106/2015, 19th February.
[41] See footnotes 13, 14, 21 and 23.

function but also contribute to its directive function, making a certain normative message clearer by putting the emphasis on certain groups.

Regarding motives, one might wonder why it is necessary to require special grounds for a conduct which already consists in fostering, promoting or inciting hatred, hostility, discrimination or violence. What is added by the requirement that such a conduct be carried out for racist reasons or any other reasons? The explanation is that this requirement acts as a filter, leaving only conducts carried out for these reasons, and not others (filtering out non-criminal conduct such as, for example, envy or hatred of the wealthy). Similarly, the law, in some authors' opinions, seems to extend its protection only to certain groups of people considered "target groups" (due to race, nationality, religion, etc.).

With respect to this selective protection, another key question is whether the offence requires that the people targeted belong to a particularly vulnerable group. The Spanish Supreme Court has directly referred to this, until recently, as a *sine qua non* requirement: hate crimes must be directed against groups with vulnerable characteristics.[42] Nevertheless, in a recent judgement, the Spanish Supreme Court concluded that these offences do not require the victims to be vulnerable. The Court expressly relies on the fact that the legislator would have stated this if he had wished to do so, as well as on the fact that the legal provision protects all citizens as long as the person or persons affected fit into one of the grounds of discrimination exhaustively established by the legislator, whether or not they were minorities, vulnerable or disadvantaged.[43] In my view, none of these reasons are sufficient to reach that conclusion.

Once again, the ambiguous wording of the law creates problems with respect to the question of vulnerability. Here the legal wording is not sufficiently clear, which creates challenges when deciding, in practice, whether it includes the protection of groups frequently exposed to hate speech, such as the police.[44] Any interpretation and subsequent conclusion must be made by using historical criteria (for instance, taking into account the background: reasons for the inclusion of these crimes in the law), logical criteria (e.g. observing the type of motives that are included which are generally characteristic of those who attack vulnerable

---

[42] See, for instance, STS 47/2019, 4th February; STS 185/2019, 2th April and STS 646/2018, 14th December. A clarifying summary of the Spanish and European legal practice on these offences can be found at: https://libex.es/ (accessed 9 February 2023).
[43] STS 437/2022, 4th May. This judgement extends the same interpretation to the same question raised by section 22.4.
[44] See the judgement on the so called "Alsasua case" STS 2/2019, 7th March, in which the claim that hate crimes could protect the police was dismissed. Similarly, ECHR Savva Terentyev v. Russia, Application No. 10692/09, 28th August 2018.

groups), systematic criteria (observing the alternatives offered by the law to punish harm caused to members of other groups) or teleological criteria (for instance, considering that only the aim of protecting vulnerable groups gives meaning to the section).

Another unsolved question, related to the above, is whether relevant hate speech is only speech directed against persons who belong to minority groups. Again, the law does not specify this, nor is the question resolved by taking a position on the requirement of the vulnerability (or not) of the group to which the victim belongs. Even though it would seem that, in order to be considered vulnerable, groups are required to be minorities, this is not always the case—an obvious example is the case of women (covered by the legal references to sex or gender). Although some of the legal grounds of discrimination might indeed be incompatible with the minority character of a group, the characteristic of minority is required by some courts in Spain. However, to reconcile the apparent inconsistency of including women in this category, one must be aware that the quality of "minority" does not rest only on quantitative reasons, but sometimes on qualitative ones. That is to say, a group can be considered a minority due to its minority weight in social participation, which is the case of women.[45] This is a good example of the use of a teleological interpretation, since it takes into account the purpose of the legal regulation in question in order to reach this conclusion.

4.2.3 Limits imposed by the right to freedom of expression

The criminalisation of hate speech and, of course, the interpretation and application of criminal law by judges, cannot ignore the need to respect the freedom of expression. In Spain, freedom of expression constitutes a fundamental right that enjoys the highest constitutional protection. Freedom of expression is widely recognised by the Spanish courts not only as a fundamental right, but also as a right with a preferential position, due to its role in ensuring political pluralism in a democratic state.[46] Indeed, freedom of expression provides ample scope for the free expression of ideas and opinions and, of course, for criticism. The idea is, as the Spanish Constitutional Court has stated, that an expression must be tolerated "even when it is harsh and may annoy, disturb or upset the person to whom it is addressed, since this is required by pluralism, tolerance and the spirit of openness,

---

[45] STSJ Catalunya (Judgement of the High Court of the Autonomous Region of Catalonia) 142/2020, 22nd June.
[46] Thus, among others, Judgement of the Constitutional Court (hereinafter referred to as STC) 174/2006, 5th June.

without which there is no democratic society". And even expressions that "attack the democratic system itself" must be respected because "the Constitution also protects those who deny it".[47] The first statement is in line with the doctrine of the European Court of Human Rights.[48] However, the rulings of this Court, which do not support a militant model of democracy, show a greater restriction of freedom of expression than the rulings of the Spanish Constitutional Court (Rodríguez Ramos 2020: 204–205).[49]

In order to achieve a ruling, any limits imposed on freedom of speech require, in each individual case, a delicate balancing act between what is permitted and what is unacceptable. The most difficult point to agree on is what precisely is required for a conduct to be deemed intolerable. There is consensus that the limitation of free speech should only be justifiable in the case of expressions that endanger or damage another specific constitutional value of fundamental importance. Nevertheless, not all speech that is intolerable should be sanctioned specifically by criminal law.[50] Because of the severity of criminal penalties and their consequences, a criminal response should be reserved for only the most intolerable speech. That is to say, it must verifiably represent an intentional incitement to commit a crime against people's basic rights.[51]

According to the jurisprudence of the Spanish Constitutional Court, the judge must carry out the aforementioned analysis prior to the application of the criminal offence to assess whether or not the conduct in question is a manifestation of the legitimate exercise of freedom of expression.[52] Yet, even if the speech in question does not constitute a legitimate exercise of this fundamental right, and even if it is legitimately foreseen as an offence in the criminal provision, the judge cannot

---

[47] STC 176/1995, 22nd December.
[48] ECHR Handyside v. The United Kingdom, Application No. 5493/72, 7th December 1976; Castells v. Spain, Application No. 11798/85, 23rd April 1992; Fuentes Bobo v. Spain, Application No. 39293/98, 29th February 2000; Erbakan v. Turkey, Application No. 59405/00, 6th July 2006, among other judgements.
[49] For the position of the Spanish Constitutional Court on this point see STC 235/2007, 7th November.
[50] STC 259/2011, 12th April; STC 35/2020, 25th February.
[51] A quite similar—but not identical— proposal for an alternative criminal regulation has been promoted in Spain by the Criminal Policy Study Group (Grupo de Estudios de Política Criminal 2019: 34). The Criminal Policy Study Group is an association made up of criminal law professors, judges and prosecutors, which works to contribute to a more rational criminal policy by developing alternative proposals for the regulation of various aspects of criminal law in Spain. More information available at: https://politicacriminal.es (accessed 9 February 2023).
[52] STC 35/2020, 25th February.

react disproportionately to the act of expression when applying the criminal norm.[53]

However, as I have pointed out, freedom of expression is not only a limit for judges when assessing a case, it also represents a limit for the formulation of laws by the lawmaker at an earlier stage. From this perspective, many have severely critiqued the broadness of the formulation of hate speech crimes in Spain.

5 Conclusions

The questions which I have presented here, raised by the legal status of hate crimes in Spain, have shown the reader that numerous challenges stem from the wording of Spanish hate crime laws. In Spanish criminal law, the commission of common offences for reasons of hatred or discrimination results in an aggravation of responsibility. In addition, there are hate speech crimes, the most representative of which consist in "fostering, promoting or inciting" hatred or discrimination.

With regard to both types of offences, the requirement that the perpetrator act for certain motives raises serious difficulties of justification and proof. This legal requirement makes these offences particularly unstable when it comes to their interpretation and judicial application. Concerning proof, difficulties are considerable given that criminal punishment depends on a motive of discrimination or hatred that needs to be verified through the analysis of the language used, the circumstances of the discourse and those others regarding the accused. The difficulties also lie in the vagueness of the expressions that the law uses when describing these criminal behaviours, particularly in the case of hate speech crimes.

I have argued that limiting the interpretation of an expression as a hate speech crime to only those manifestations which are severe is of paramount importance, in order to contain the literal scope of offences that are too broadly or unclearly formulated in the law. As has been shown, careful interpretation is required to clarify the meaning of these ambiguously formulated laws. This task requires careful analysis of the laws' exact wording, in the light of fundamental constitutional principles and technical criteria of criminal law.

With regard to hate speech crimes, it should also be stressed that only external conduct can cause harm to others. Thus, these facts should constitute the minimum condition necessary for criminal law to act. "Words" (by which I mean written or spoken words) are also facts, but the harm that can be caused by

---

[53] STC 110/2000, 5th May.

words is generally less than that caused by actions, such as physical violence. In many cases the role of criminal law in dealing with the harmful use of words is therefore particularly difficult to define.

In fact, the words considered to be most serious are not only those that constitute threats, but also those that under the Spanish Criminal Code make up the so-called "punishable preparatory acts". That is, acts aimed at committing specific crimes, such as incitement to commit a crime. Under Spanish law, such acts are only punishable as attempts to get others to join a criminal initiative proposed by the speaker.

In my view, these acts are precisely what should be punishable as so-called "hate speech crimes". I believe that this teleological explanation allows us to avoid the protection of vague legal rights, such as equality or non-discrimination, which lack the objective conditions that a legal interest protected by criminal law should possess. Moreover, the protection of vague legal rights can go awry, leading to the criminal punishment of conducts of dubious harm. A teleological explanation requires that the potentially dangerous effect of the message be materially provable and proven at court. As opposed to the above-mentioned vague rights, interpreting hate speech in this way brings to the surface fundamental human rights, such as the right to life, physical integrity, dignity or freedom.

Any hate speech can indeed be very dangerous, both because it contributes to the undermining of the social position of the targeted groups, or to the hindering of their progress, and, above all, because it can incite violence against them. However, in view of the current regulation, an attempt is often made to justify the sanction of criminal law by considering effects that *could* result from the various conducts, hypothetically carried out afterwards by third parties (e.g. "what if everybody did it?"), rather than focusing on the effect of the actual conduct being punished. This leads inevitably to a distortion in the proportionality between the penalty and the offence, while the imputation of guilt to a specific perpetrator implies the attribution of a responsibility which might not constitute a criminal offence. As the Spanish Constitutional Court has said, "between hatred that incites the commission of crimes, hatred that sows the seeds of confrontation and erodes the essential values of coexistence, and hatred that is identified with animosity and resentment, there are slight differences that cannot be ignored by the criminal judge on the basis that everything that is not accepted in freedom of expression is intolerable and therefore necessarily criminal".[54]

What we have been considering here is directly linked with the justification of the use of criminal law against hate speech. However, as I have also mentioned,

---

54 STC 235/2007, 7th November.

once the use of criminal law is chosen to avoid certain kinds of conduct it is the law, in accordance with our legal system, that must establish what is prohibited.

As we have seen, however, the scope of the prohibited conducts is not sufficiently clear. In the interpretation of criminal law, we must first heed the meaning of the words of the law. However, in the case of hate speech offences, the importance of legal wording must be relativised to a greater extent, because its function of guiding the personal autonomy of citizens needs to be carried out within the limits set by the legal system as a whole. As also mentioned earlier, there are external limits to criminal law (some of an axiological nature which establish conditions of validity or others such as the right to freedom of expression, imposed by the Constitution) and internal limits established by criminal law itself (either of a logical, systematic or teleological nature). There are also limits of a technical nature (in other words, limits imposed by criminal policy, such as the restrictions imposed by preventive needs; for instance, when having to decide whether what is required for the crime to be sanctioned is the result of hate speech or only the expression of hate).

The task of determining the legal meaning (interpretation) should remain within the margins of the "possible literal wording" of the legal terms, as the principle of legality requires. But, as we have seen, this is not very helpful here, due to the ambiguity of the legal text. In a state governed by the rule of law, it is absolutely essential that the words of the law themselves adequately delimit the acts subject to criminal punishment. As the ECRI has stated, without such clarity and precision, there is likely to be an absence of legal certainty as to the scope of the prohibited conduct (Explanatory Memorandum to Recommendation No. 15 On Combating Hate Speech, 2015, No. 175). Clarity and precision are essential both for political reasons (principally, to ensure legal certainty and the freedom of citizens, but also to reinforce community support) and for technical or preventive reasons (these preventive reasons not only refer to general deterrence of the prohibited conduct, but, also and again, to the public support generated by the rules of criminal law). Sometimes, and this is the case with hate speech crimes in Spain and in other countries, the current criminal regulation is not in a position to offer these guarantees.

The only alternative regulation that would respect the constitutional and technical limits mentioned above is one which limits the scope of the intervention of criminal law to cases in which there is a public and direct incitement to commit criminal offences against people's basic rights.

References

Arroyo de las Heras, Alfonso. 1997. Los motivos racistas, antisemitas o discriminatorios como circunstancia agravante. In Javier Muñoz Cuesta (ed.), *Las circunstancias agravantes en el Código Penal de 1995*, 107–124. Pamplona: Aranzadi.
Atienza, Manuel. 2017. *Introducción al Derecho*. México City: Fontamara.
Borja Jiménez, Emiliano. 1999. *Violencia y Criminalidad racista en Europa occidental: La respuesta del Derecho Penal*. Granada: Comares.
Borraz, Marta. 2021. "El asesinato de Samuel desata una ola de protestas contra la violencia hacia las personas LGTBI" elDiario.es, 5 July 2021. https://www.eldiario.es (accessed 9 February 2023).
Cuerda Arnau, María Luisa. 1996. In T.S. Vives Antón (ed.). *Comentarios al Código Penal de 1995*, 238–247. Valencia: Tirant lo Blanch.
Fuentes Osorio, Juan Luis. 2017. El odio como delito. *Revista Electrónica de Ciencia Penal y Criminología* 19–27. 1–52.
Goyena Huerta, Jaime. 2015. De las circunstancias que agravan la responsabilidad criminal. In Manuel Gómez Tomillo (ed.), *Comentarios prácticos al Código Penal*, t. I, 350–355. Pamplona: Aranzadi.
Grupo de Estudios de Política Criminal. 2019. *Una propuesta alternativa de regulación de los delitos de expresión*. Valencia: Tirant lo Blanch.
Harel, Alon & Gideon Parchomovsky. 1999. On hate and equality. *The Yale Law Journal* 109(3). 507–539.
Hortal Ibarra, Juan Carlos. 2012. La circunstancia agravante por motivos racistas o discriminatorios: Una propuesta restrictiva de interpretación. *Cuadernos de Política Criminal* 108(3). 31–66.
Lara, Rafael. 2008. ¿Regulación de Flujos? 20 años de muerte en las fronteras. *Asociación Pro Derechos Humanos de Andalucía*. Derechos Humanos en la Frontera Sur. http://www.apdha.org (accessed 9 February 2023).
Laurenzo Copello, Patricia. 1996. La discriminación en el Código Penal del 95. *Estudios Penales y Criminológicos* 19. 219–288.
Laurenzo Copello, Patricia. 2019. La manipulación de los delitos de odio. In Esther Pomares Cintas & Juan L. Fuentes Osorio (eds.), *Un juez para la democracia: Libro homenaje a Perfecto Andrés Ibáñez*, 453–468. Madrid: Dykinson.
Peiró, Patricia. 2021. "La justicia avala el cartel electoral de Vox contra los menores extranjeros: Son un evidente problema social y político". *El País*, 5 July 2021. https://elpais.com (accessed 9 February 2023).
Pereda, Carlos, Walter Actis & Miguel Ángel De Prada. 2003. La sociedad española y la inmigración extranjera. *Papeles de Economía Española* 98. 16–31.
Río Ruiz, Manuel Ángel. 2003. *Violencia étnica y destierro: Dinámicas de cuatro disturbios antigitanos en Andalucía*. Granada: Maristan y Fundación Secretariado Gitano.
Rodríguez Ramos, Marta. 2020. España y Europa frente al discurso del odio: Una aproximación comparativa a los límites a la libertad de expresión en la jurisprudencia española y en la del Tribunal Europeo de Derechos Humanos. *Revista Penal* 46. 190–206.
Sistare, Christine. 1987. Agent motives and the criminal law. *Social Theory and Practice* 13(39). 303–326.
Solan, Lawrence M. & Peter M. Tiersma. 2018. *Hablar sobre el delito: El lenguaje de la justicia penal*. Madrid: Marcial Pons.
Steiker, Carol. 1999. Punishing hateful motives: Old wine in a new bottle revives calls for prohibition. *Michigan Law Review* 97(6). 1857–1873.

Legal sources

1 Legal instruments

Spain
LO 4/1995 Amending the Criminal Code for Punishing the Direct and Public Incitement to Commit Genocide.
LO 1/2015 amending LO 10/1995 of the Criminal Code.
LO 8/2021 on the Full Protection of Children and Adolescents against Violence.
LO 6/2022 on the Full Protection for Equal Treatment and Non-Discrimination.

European Union
Council Framework Decision 2008/913/JHA of 28th November 2008 On combating certain forms and expressions of racism and xenophobia by means of criminal law, OJ L 328, 6th December 2008.

Council of Europe
ECRI General Policy Recommendation No. 1 of 1996, on Combating Racism, Xenophobia, Antisemitism and Intolerance. Adopted on 4th October 1996.
ECRI General Policy Recommendation No. 7 of 2002 on National Legislation to Combat Racism. Amended on 7th December 1017.
ECRI General Policy Recommendation No. 15 of 2015 on Combating Hate Speech. Adopted on 8th December 2015.

United Nations
Annual report of the United Nations High Commissioner for Human Rights and reports of the Office of the High Commissioner and the Secretary-General (A/HRC/22/17/Add.4)(Rabat Plan of Action).

2 Cases

Spain
a) Constitutional Court
STC 176/1995, 22nd December.
STC 110/2000, 5th May.
STC 174/2006, 5th June.
STC 235/2007, 7th November.

STC 259/2011, 12th April.
STC 35/2020, 25th February.

b) Supreme Court
STS 259/2011, 12th April.
STS 299/2011, 25th April.
STS 646/2018, 14th December.
STS 106/2015, 19th February.
STS 47/2019, 4th February.
STS 2/2019, 7th March.
STS 185/2019, 2nd April.
STS 675/2020, 11th January.
ATS 16th November 2020.
STS 437/2022, 4th May.

c) Provincial Criminal Courts
SAP Burgos 420/2001, 20th October.
SAP Barcelona 713/2014, 22nd July.
SAP Barcelona 27th April 2018.
SAP Barcelona 392/2018, 14th May.
SAP Barcelona 702/2018, 8th November.
SAP Barcelona 86/2020, 5th February.
AAP Madrid 631/2021, 1st July.
SAP Santa Cruz de Tenerife 107/2014, 7th March.
SAP Tarragona 455/2019, 8th November.
SAP Valladolid 56/2019, 8th March.

d) High Court of the Autonomous Regions
STSJ Catalunya 142/2020, 22nd June.

Council of Europe (European Court of Human Rights)
ECHR *Angelova and Iliev v. Bulgaria*, Application No. 55523/00, 26.7.2007 [Section V].
ECHR *Féret v. Belgium*, Application No. 15615/07, 16th July 2009.
ECHR *Savva Terentyev v. Russia*, Application No. 10692/09, 28th August 2018.
ECHR *Handyside v. The United Kingdom*, Application No. 5493/72, 7th December 1976.
ECHR *Castells v. Spain*, Application No. 11798/85, 23rd April 1992.
ECHR *Fuentes Bobo v. Spain*, Application No. 39293/98, 29th February 2000.
ECHR *Erbakan v. Turkey*, Application No. 59405/00, 6th July 2006.

USA
Supreme Court
Wisconsin v. Mitchell, 508 US 476 (1993).
Schenck v. United States, 249 U.S. 47 (1919).

3 Other documents

Ministry of Interior (Spain). Annual Series. Hate crimes. Victimisations by cause of hate crimes, available at: https://estadisticasdecriminalidad.ses.mir.es (accessed 9 February 2023).

Ministry of Justice of the Kingdom of Spain. State Public Prosecutor's Office in its Guidelines for interpreting hate crimes under section 510 of the Criminal Code (2019). https://www.mjusticia.gob.es (accessed 9 February 2023).

LibEx. Grupo de Trabajo sobre Libertad de Expresión. https://libex.es/ (accessed 9 February 2023).

Subject index

affect (effect) 7, 125, 152, 160, 207 fn14
affect (emotion evaluation) 1
affective rhetoric 160
aggravating circumstance 5, 6, 117, 118, 140, 146, 157, 184, 203–212, 224
alien 19, 34, 63–76, 79, 80, 81, 82, 85, 86, 88, 96, 97, 104, 107, 108, 109, 142
aloneness 8, 58
ambiguity, vagueness 28, 52, 71, 76, 78, 125, 127, 136, 137, 144, 190, 201, 208, 211, 214, 217, 218, 219, 221, 224, 225, 226
analogy *in malam partem* 207
animus iniuriandi ('intent to offend') 216
animus (or hostility) model 6
antimigration 1, 2, 3
anti-Roma 207, 207 fn13, 207 fn14, 212 fn21, 213, 214 fn23, 220
anti-Semitism 149, 157, 161, 164, 166, 167, 189 fn21, 204, 213, 220
aporophobia 207, 207 fn13, 213, 214 fn23
appraisal theory 1
argumentative rhetoric 131, 160
Arlington Heights (1977) 85, 85 fn1, 86, 89, 90, 91, 93, 96, 111, 113
ascriptivist 6, 17
attitude, attitudinal 1, 2, 3, 7, 10, 15, 16, 26, 56, 85, 103, 121, 131, 138, 143, 144, 153, 154, 173, 186, 187, 191 fn22, 209
Australian Criminal Code 181, 181 fn13

bilingualism 51, 52, 53, 54
blameworthiness 209, 210, 210 fn17, 214
borders (national) 2, 9, 36 fn8, 40, 40 fn16, 41, 42, 43, 69, 82, 85, 88, 162, 163, 164, 174
Brandenburg test 8, 179 fn9, 190, 191

Canadian Criminal Code 179, 180
challenging question 128, 130, 131, 144
Civil law 13, 173, 177, 179, 182, 183, 184, 196
cognitive, cognitivist, non-cognitivist 6, 14, 17, 93, 175, 182
Common law 13, 173, 177, 179
Complex speech act 174, 178, 189, 191, 192, 195, 196
condescension 123, 138, 144

consequential harm 153
conservative, conservatives 19, 55, 63, 68, 69, 72–76, 78, 81, 82
conspiracy theories 20, 39 fn14, 149, 154, 157–161, 164–169
constitutional principles 202, 224
constitutive harm 153, 154
Council of Europe 183, 183 fn14, 204, 228, 229
– see also European Court of Human Rights
covert disparate treatment 3
covert hate speech 20, 144, 149, 150, 153, 154, 154 fn3, 155, 155 fn4, 156, 157, 159, 161, 164, 169, 194
covert hatred 3, 156
covert Islamophobia 149, 161, 169
covert anti-Semitism 166, 169
criminal alien 71
criminal law 5, 6, 8, 10, 29, 90, 105, 202–205, 205 fn7, 206, 209, 210, 211, 214, 219, 220, 223, 223 fn51, 224, 225, 226, 228
criminal law of facts 209
criminology 24, 25, 29, 34, 43
crimmigration 25, 39
crisis 2, 3, 23, 25, 44, 162, 173, 187

danger 2, 8, 18, 23, 25–44, 31 fn4, 65, 69, 71, 74, 105, 111, 142, 152, 156, 157, 169, 174, 174 fn1, 176, 178, 183, 194 fn29, 204 fn3, 207 fn13, 214, 219, 219 fn35, 219 fn36, 220, 223, 225
dangerousness 23, 25, 29–35, 37, 38, 39, 42, 43, 44
democracy 33, 129, 223
defamation 11, 12, 155
descriptivist 6, 17
detention 41, 67, 71–75, 81, 175
dialect 58
discourse analysis 1, 113, 117, 145, 160, 185, 189
discrimination 3, 5, 6, 8, 65, 89, 89 fn3, 102, 104, 112, 113, 150–154, 156, 175–179, 175 fn4, 181, 182, 184, 185, 189, 195, 196, 203, 204 fn5, 205–208, 208 fn13, 210–214, 216, 218, 219, 219 fn37, 220, 220 fn39, 221, 222, 224, 225
discriminatory intent 89, 90, 92

dismissals 123, 127, 128, 138, 139, 140, 144
disparagement 3, 4

economic situation 3, 4, 31 fn4, 35, 37, 38, 39, 40, 41, 64, 109, 111, 174, 203
ECRI's General Policy Recommendation No.15 182
election campaign 216, 217
emotion 1–5, 1 fn1, 8, 9, 11, 56, 118, 120, 121, 124, 135, 152, 156, 160, 167, 169, 174, 174 fn2, 175, 179
emotional stress 4, 179
epistemic authority 149, 152–155, 157, 163, 168, 169
equal protection, participation, treatment 6, 8, 88, 89, 89 fn3, 90, 118, 153, 202, 207 fn13
equality 4, 8, 118, 149 fn1, 202, 210, 214, 225
etymology 18, 23, 24–27
European Commission against Racism and Intolerance (ECRI) 182, 182 fn14, 184, 204, 205, 205 fn8, 226
European Court of Human Rights (ECHR) 7, 153 fn2, 192, 192 fn24, 193, 194, 204, 205, 205 fn6, 205 fn9, 219, 221 fn44, 223, 223 fn48, 229
European law 8, 173, 177, 182, 195, 201, 204
European Union (EU) 66, 132, 134, 140, 149, 149 fn1, 151, 153, 154, 161, 167, 204, 205, 215
evidence (legal) 12, 13, 16–19, 43, 55, 57, 135, 210, 219

facially neutral 89, 89 fn3, 92, 94, 111, 112
fallacies 3, 163
fighting words 8
foreign accent 51, 51 fn3, 52, 53
freedom of expression 6, 7, 155, 157, 177, 179, 183, 194, 194 fn29, 195, 201, 222–225, 226
freedom of opinion and expression 177, 180, 184
fundamental right 4, 89 fn3, 176, 222, 223

German Criminal Code 183, 183 fn18, 184
Great Replacement Theory 161, 164, 169
global act 189, 195, 196
group defamation 179, 190
group harassment 186, 188, 189, 190, 193, 195

harassment 155, 181, 182, 189
harm 3, 4, 4 fn2, 5, 6, 8, 26, 27, 28, 34, 35, 37, 38, 39, 39 fn13, 43, 75, 121, 140, 149, 151–154, 156, 176–179, 194 fn29, 206, 208, 210, 213, 214, 220, 222, 224, 225
hate crime 5, 6, 15, 87, 117, 118, 176, 178, 202–206, 206 fn11, 211 fn18, 217, 219 fn37, 221, 221 fn44, 224
hate speech 2–4, 4 fn2, 5–9, 11–20, 39 fn13, 65, 117–121, 124–128, 135, 140, 143–146, 149–155, 154 fn3, 155 fn4, 156–160, 164, 169, 173–196, 194 fn27, 194 fn28, 201, 204 fn5, 205, 206, 212, 214, 215, 216, 218, 219 fn35, 221, 222, 224, 225, 226
hate speech crimes 6, 13, 15, 151, 157, 178, 201, 206, 212, 214, 218, 224, 225, 226
hatred advocacy 186, 187, 193, 195
hatred dissemination 186, 187, 193, 195, 219 fn35
hatred on sexual orientation grounds 176, 180, 180 fn11, 182, 184, 212, 213, 214
human rights 6, 7, 8, 36 fn9, 41, 43, 44, 125, 176, 177, 178, 180, 192, 225

ideology 32, 63, 64, 68, 72, 74, 75, 77, 78, 184, 185, 196, 203, 204, 207, 212, 213
illegal alien 63–66, 69, 71, 72, 74, 81
illocutionary act 11, 14, 15, 16, 19, 149, 153, 157, 168, 179, 181, 183, 185, 188, 190–195
illocutionary force 16, 19, 149, 153, 157, 168, 191, 194
ill-wishes 123, 134, 138, 139, 143, 144
immigration 1, 8, 19, 25, 39, 58, 65, 66, 67, 71, 73, 81, 85, 86, 88, 89, 90, 96, 97, 102–109, 113, 131, 141, 155, 159–163, 162 fn10, 168, 202, 203, 204
imminence test 8, 124, 179, 191, 194 fn29
impoliteness 19, 117–130, 132, 133 fn8, 134, 135, 136, 138, 140–146
impotent despair 8
incitement 3, 8, 11, 16, 17, 128, 150, 151, 152, 156, 176–184, 183 fn15, 183 fn17, 186–195, 204 fn4, 205, 208, 212, 216–219, 223, 225, 226
incitement of masses 32, 34, 183, 183 fn15
incitement to hatred 16, 177, 179, 180, 183, 183 fn17, 184, 186, 187, 188, 191–195, 205, 212, 216–219
inferiority 4, 102, 111
inflammatory speech 8, 18

insult 3, 5, 7, 12, 120, 122, 127, 128, 130, 132, 133, 133 fn8, 136, 137, 138, 144, 145, 154, 158, 160, 173, 181, 182, 183, 188, 189, 208
intention 14–18, 27, 42, 65, 91, 93, 112, 114, 120, 121, 134 fn9, 149, 151, 156, 161, 161 fn8, 178, 179, 183, 186, 187, 191 fn22, 194, 194 fn29, 195, 196, 216, 217, 223
intentional infliction of emotional distress 179
interdiscursivity 87, 92, 93, 112
International Covenant on Civil and Political Rights 5, 150, 177, 178, 180, 192
International Convention on the Elimination of All Forms of Racial Discrimination (ICERD) 5, 150, 177, 178, 179
International law 1, 150, 173, 177
interpretation 6, 10, 10 fn7, 11, 12, 15, 17, 24, 28, 31, 37, 43, 79, 86, 91, 99, 112, 113, 118, 123, 136, 173, 194, 195, 196, 201, 202, 207, 210, 212, 214, 217, 219, 221, 221 fn43, 222, 224, 226
Islamophobia 149, 161

labelling 24, 34, 35, 38, 39, 44, 127, 153
Language Analysis for the Determination of Origin (LADO) 49, 50, 54, 55, 57, 58
language attrition 49, 50, 54, 56, 57
language of the law 85, 86, 87, 89 fn3, 91, 92, 93, 111, 112
lawmaker 92, 105, 224
legal approaches to hate speech prohibition 173, 176, 178, 185, 187, 195
legal meaning 10, 86, 87, 91, 92, 112, 113, 154, 173, 177, 195, 196, 202, 210, 222, 224, 226
legislative history 90, 93, 96, 103
legislative record 88, 89, 90, 92, 94, 96, 111
liberal 64, 68, 70, 72–76, 81, 154, 209
Library of Congress 64, 65, 66
life chances 4
limitation of personal freedom 4
literal meaning 10, 14, 17
literalness 15, 17
locutionary act 11–16

macrostructure 186, 189, 190, 195
Member State law 173, 177, 183, 195
message enforcer 123, 144
migration 1, 2, 3, 8, 9, 12, 19, 23, 25, 35–39, 36 fn8, 36 fn9, 40 fn16, 41–44, 52, 56, 58, 65, 66, 67, 71, 73, 81, 85, 86, 88, 89, 90, 96, 97, 102–109, 113, 131, 141, 149, 155, 156, 157, 159–163–169, 162 fn10, 173, 174, 202, 203, 203 fn1, 203 fn2, 204
minority group 5, 6, 8, 196, 211, 222
motives 6, 10 fn8, 87, 92, 93, 206–209, 210, 210 fn17, 211, 212, 216, 220, 221, 224
multilingualism 50

negative emotions 2, 3, 8, 160, 174, 175
New World Order (NWO) 158–161, 164, 167, 168, 169
no-imminence standard 175
non bis in idem ('prohibition of double jeopardy') 206
noncitizen 63–68, 70–76, 78, 81, 82, 104

offence (politeness) 118, 121, 124, 185
offence (criminal) 7, 151, 180, 181, 184 fn15, 195, 203–206, 206 fn11, 207 fn14, 208–215, 218–221, 221 fn42, 223–226
online comments 117, 118, 124, 127, 128, 132, 137, 138, 142, 143, 145, 150, 163
oral arguments/discourse 64, 67, 68, 70, 71, 73, 74, 76, 81
overt disparate treatment 3
overt hate speech 10, 44, 149, 150, 154, 155, 156, 163, 167, 168, 169

perlocutionary act 16, 149, 153, 169, 179, 180, 181, 183, 185, 188, 190
perlocutionary effect 149, 153, 169, 180, 181, 183
personal security 2, 4
pluralism 7, 222
pointed criticism 122, 127, 128, 133, 133 fn8, 136, 137, 145
power (legislative, legal, political) 41, 42, 57, 74, 92, 96, 101, 174
power (of persuasion, language) 3, 33, 35, 39, 44, 82, 124, 138, 142, 149, 158, 160
power (social power) 27, 28, 31, 32, 33, 34, 38, 39, 40, 41, 42, 43, 44, 74, 92, 110, 111, 122, 124, 138, 142, 158, 160, 188
practical authority 152–157, 163
presupposition 123, 127, 128, 129, 143, 144
principle of legality 9, 10, 207, 226

psychic destruction 4
Public Order Act 180, 180 fn11, 181, 181 fn12

qualifiers 150, 164, 165
qualifying circumstance 206, 209, 210, 211

Rabat Plan of Action 124, 194 fn29, 220, 220 fn39
racial animosity 181, 182, 189
racial animus 64, 85, 87, 89, 93–96, 98, 102, 108, 111
racial harassment 182
racial hatred 19, 106, 150, 178, 180, 181, 182, 193, 208
racial insult 3, 5, 7, 12, 120, 122, 127, 128, 130, 132, 133, 133 fn8, 136, 137, 138, 144, 146, 154, 160, 173, 181, 182, 183, 188, 189, 208
Racial Vilification Act 182, 187
racism 41, 43, 109, 154, 176, 184, 185, 189 fn21, 193, 201, 203, 204, 204 fn5, 205, 208, 212, 213, 216
racist grounds 211
re-entry 67, 86, 104, 108
refugee 1–5, 8, 43, 49, 50, 55, 56, 125, 134, 135, 136, 160, 162–166, 162 fn10, 166 fn11, 167, 168, 175, 176, 215
religious hatred 117, 118, 140, 151, 157, 176, 178, 180, 180 fn11, 182–185, 192, 203–208, 212, 213, 220, 221
reproach 209, 214
rhetorical questions 130, 131, 158, 164, 165, 166
risk 2, 26, 28, 29, 30, 34, 37, 38, 41, 44, 144, 151, 176, 177, 189, 193, 195, 220, 220 fn39
rule of law 5, 19, 210, 226

selective discrimination model 6
self-esteem 4
shame 2, 4
silencer 123, 128, 138, 142, 143, 144
social exclusion 189, 190, 195, 196, 207, 207 fn13
social promotion 4
social threat 18, 174, 174 fn1
Spanish Criminal Code 184, 201, 202, 204, 204 fn4, 205 fn10, 206, 206 fn11, 207, 207 fn14, 211, 211 fn18, 212, 218, 218 fn34, 225, 228, 230
stateless person 8, 66, 67, 80
stereotype 38, 149, 152, 154, 157, 164, 175

stereotype threat 4
super goal 186, 189, 190, 195, 196
superstructure 186, 189, 190, 195
Supreme Court 19, 36 fn8, 63, 67, 81, 82, 85, 85 fn1, 88–91, 89 fn3, 94, 102, 108 fn4, 190, 191, 193, 211, 212, 212 fn20, 219, 220, 221

technical legal definition of hate speech 176, 177, 195
threat 2, 3, 4, 13, 18, 19, 26, 28, 29, 30, 32, 34, 35, 37, 38, 38 fn12, 39, 41, 43, 44, 86, 106, 108, 110, 111, 120–123, 128, 131, 133, 136, 138, 139, 140–144, 140 fn11, 149, 152, 153, 156, 157, 161, 168, 174, 181, 182, 188–191, 191 fn22, 206, 208, 225
Tort law 179
trial 26, 71, 85, 95, 101, 219

unaccompanied foreign minors 217
undocumented 36 fn8, 36 fn9, 66, 104, 105, 107, 108
unfairness 8, 57
United States Code (U.S.C.) 85–87, 90, 92, 93, 103, 105, 111, 113
United States Constitution 8, 179
Universal Declaration of Human Rights (UDHR) 177, 178, 180
US Naturalization Act 64

vagueness 201, 218, 224
verbal attacks 10, 14, 117, 120, 121, 166, 206, 209, 217
victim 4, 4 fn2, 6, 7, 8, 28, 31, 120, 123, 140, 152, 157, 176, 184, 196, 203, 206 fn12, 207–210, 210 fn17, 211, 215, 221, 222
vocative 122, 132, 135, 136, 138, 144
vulnerable groups 3, 5, 11, 27, 35, 36, 38, 39, 40, 44, 44 fn17, 117, 151, 163, 173–176, 179, 180 fn11, 187, 195, 210, 211, 212, 221, 222

women 64, 101, 107, 129, 130, 162, 163, 222
wrongfulness 209, 210

xenophobia 44, 154, 193, 201, 202, 204, 205, 208, 212, 216
xenophobic 5, 43, 44, 149, 203
xenophobic grounds 203–209, 212, 219

www.ingramcontent.com/pod-product-compliance
Lightning Source LLC
Chambersburg PA
CBHW050522170426
43201CB00013B/2054